Thomas Butts
Michigan State University

Problem Solving in Mathematics

ELEMENTARY NUMBER THEORY AND ARITHMETIC

Scott, Foresman and Company
Glenview, Illinois London

Cover photograph by Wm. Franklin McMahon

Library of Congress Catalog Card Number 72-90234
ISBN 0-673-07723-3
AMS Classification Number 98B10

Printed in the United States of America.
Regional offices of Scott, Foresman and Company
are located in Dallas,
Oakland, N.J.,
Palo Alto,
and Tucker, Ga.

Preface

Why do you like mathematics? For me, and I suspect for most people who like mathematics, the one most appealing feature is the joy of solving a problem—the sense of satisfaction which comes after guessing the solution and then proving it. Unfortunately, most prospective mathematics students are denied this pleasure in their early collegiate mathematical careers. Most textbook problems at this level are either skill problems or "prove that" problems. While both types of problems have their place, they both ignore the creative aspect of mathematics.

The purpose of this book, then, is to introduce the joy of problem solving to students early in their mathematical careers. It is largely a collection of problems dealing with arithmetic and elementary number theory, most of which are phrased: "For what (type of) integers n which . . . .", or "Can ___ always be ___ ?" In addition, there is some discussion of problem solving from both the student's and the teacher's point of view. The only *formal* background necessary to solve almost all of the problems is one year of high school algebra. In a few problems, concepts from second-year algebra are used.

This format should be appealing for several types of courses: pre-service and in-service courses for elementary and secondary teachers, a non-calculus course to introduce students to mathematical reasoning, and summer institute courses in problem solving. I have used this material with four groups of pre-service elementary and junior high teachers and a few secondary teachers.

Chapters 1–8 form the core of the book and should be covered in order; Chapters 9–19 are largely independent, with the exceptions that Chapter 9 should precede Chapter 10, and Chapter 15 should precede Chapter 16. Hints and Comments are at the end of each chapter. Hints are given to problems which I feel are crucial to the understanding of the concepts; consequently, many difficult problems have no hints. Also, there are fewer hints towards the end of the book.

At this point I would like to express my gratitude to four people: George Polya, whom I have never met but whose writings on problem solving have inspired me to write this book; James Smart, who carefully read the entire manuscript and offered many valuable suggestions; Arthur Hallerberg, who class tested the first half of the material; and Kathy Trebilcott, who typed most of the manuscript.

Thomas Butts

Contents

Problem Solving in Mathematics

Introduction

One of the main purposes of "modern math" is to show that the study of arithmetic can be based upon the knowledge of a few fundamental principles—closure, associativity, commutativity, distributivity, etc. That is, the study of arithmetic, as well as many other branches of mathematics, does not consist of merely memorizing a few hundred (seemingly independent) rules with the ultimate hope (seldom achieved) that one knows the right time to apply the right rule. Rather, the study of arithmetic consists of the knowledge of a few basic principles and the development of ability to acquire skills and to solve problems by combining these principles in many different ways and in many different contexts.

This approach to arithmetic, therefore, puts the emphasis on thinking and reasoning instead of memorizing. (For example, how does one proceed from these principles to determine how one divides two rational numbers?) The ability to think independently, to solve problems, becomes paramount; its importance cannot be overemphasized. As George Polya so convincingly put it (13 years before Sputnik):

> A great discovery solves a great problem but there is a grain of discovery in the solution of any problem. Your problem may be modest; but if it challenges your curiosity and brings into play your inventive faculties, and if you solve it by your own means, you may experience the tension and enjoy the triumph

> of discovery. Such experiences at a susceptible age may create a taste for mental work and leave their imprint on mind and character for a lifetime.
>
> Thus, a teacher of mathematics has a great opportunity. If he fills his allotted time with drilling his students in routine operations he kills their interest, hampers their intellectual development, and misuses his opportunity. But if he challenges the curiosity of his students by setting them problems proportionate to their knowledge, and helps them to solve their problems with stimulating questions, he may give them a taste for, and some means of, independent thinking.
>
> Also a student whose college curriculum includes some mathematics has a singular opportunity. This opportunity is lost, of course, if he regards mathematics as a subject in which he has to earn so and so much credit and which he should forget after the final examination as quickly as possible. The opportunity may be lost even if the student has some natural talent for mathematics because he, as everybody else, must discover his talents and tastes; he cannot know that he likes raspberry pie if he has never tasted raspberry pie. He may manage to find out, however, that a mathematics problem may be as much fun as a crossword puzzle, or that vigorous mental work may be an exercise as desirable as a fast game of tennis. Having tasted the pleasure in mathematics he will not forget it easily and then there is a good chance that mathematics will become something for him; a hobby, or a tool of his profession, or a great ambition.
>
> — **G. Polya, *How to Solve It* (Princeton University Press, 1945; Anchor Books, 1957). From the Preface to the First Printing.**

The purpose of this little book, then, is to give you a chance to "taste raspberry pie," a chance to develop your ability to solve problems, a chance to experience the creative nature of mathematics.

Perhaps no other subject is so susceptible to such extremes of good and bad teaching as is mathematics, and much "bad" teaching stems from a failure to bring out the excitement of creating mathematics. And a "sine qua non" for making mathematics exciting to a pupil is for the teacher to be excited about it himself; if he is not, no amount of pedagogical training will make up for the defect.

— R. L. Wilder, *The Evolution of Mathematical Concepts,* (Wiley, 1968) page 9.

I hate quotations.
Tell me what you know.

— R. Emerson

0
The First Problem

The following problem is designed to give you some flavor of what you will be doing while studying this book. Do as much of this problem as you can, and be sure to spend a reasonable amount of time on it. Don't give up right away, you can do it if you want to! After completing as much of this problem as you think you can do, look back over your work and ask yourself: "What have I learned about mathematics and about problem solving?" Finally, save the work you have done on this problem, and try the problem again after completing Chapter 11.

Problem 0 Consider the set of the first n positive integers, i.e., let $S_n = \{1, 2, 3, 4, \ldots, (n-1), n\}$. (For example, $S_4 = \{1, 2, 3, 4\}$.)

(a) Notice that S_4 can be partitioned into two disjoint subsets $A = \{1, 4\}$ and $B = \{2, 3\}$ such that:

(1) Every integer in S_4 is either in A or B ($A \cup B = S_4$ in set theory notation).

(2) No integer is in both A and B ($A \cap B = \emptyset$).

(3) No integer is used more than once.

(4) The sum of all integers in A equals the sum of all integers in B ($1 + 4 = 2 + 3$).

Is such a partition possible for S_5, S_6, S_7, S_8?

(b) In general, for what (type of) integers n can S_n be so partitioned? Explain your reasoning.

(c) Now consider $S_{15} = \{1, 2, 3, \ldots, 14, 15\}$. This time, partition S_{15} into five subsets of three elements each so that each integer appears in only one subset and the sum of the three integers in each subset is the same. Do this in at least three ways. Can this be done in a way in which $\{7, 8, 9\}$ is one of the subsets? Why?

(d) Now *you* try to formulate a problem involving partitioning one, or some, or all of the sets S_n in some way. First formulate a specific problem like (a) or (c), and then formulate a general problem like (b) [e.g., for what (type of) integer n is such and such true?]. Lastly, try to solve the problems you formulated.

SOME PRELIMINARY REMARKS ON PROBLEM SOLVING

When you begin to solve problems in any area, it is necessary to ask, "What do I have to go on?" and "What facts are considered to be common knowledge?" This question is especially important in the first six chapters of this book. In these chapters it will sometimes be necessary for you to make use of certain basic concepts before they are formally presented. This situation is unavoidable since these basic concepts are so closely interrelated, and many problems involve more than one such concept. I feel, however, that no serious difficulties should arise from this because you have undoubtedly encountered all or most of these concepts before. The purpose of these remarks, then, is to present a brief summary of those ideas which may be useful in solving certain problems in the first six chapters. You can then make use of the knowledge gained in these chapters to solve the problems in succeeding chapters. In addition, feel free to make use of any skills, concepts, etc. you have accumulated from your high school mathematics courses, especially algebra.

The most basic relationship between two positive integers is that one is a factor of the other. The integer* a is a factor of the integer b if there exists an integer k such that $k \cdot a = b$ (e.g. 6 is a factor of 30 since $5 \cdot 6 = 30$). The factor concept, culminating in the Fundamental Theorem of Arithmetic, is discussed in the first three chapters; but in order to appreciate some of the problems in those chapters, it is necessary to understand what happens if a is not a factor of b. In that case, when you divide b by a, you obtain a quotient q and a remainder r (with $r < a$) so that $b = aq + r$ (e.g. if $b = 37$ and $a = 14$, then $q = 2$ and $r = 9$, so that $37 = (14 \cdot 2) + 9$. This fact, known as the *Division Algorithm* (see Chapter 6), is the basis for several well-known facts which can be used in solving some of the problems in the first three chapters.

Suppose we divide every integer (one at a time) by 2. Then the remainder is either 0 or 1, so that every integer can be written in the form $2q$ or $2q + 1$; i.e. all even integers have the form $2q$ and all odd integers have the form $2q + 1$, where q is some positive integer or 0. These representations can be used to prove many familiar facts:

(1) The sum or difference of two even integers is even.
(2) The sum or difference of two odd integers is even.
(3) The product of two even integers is even.
(4) The product of two odd integers is odd.
(5) In every pair of two consecutive integers, one is even and the other is odd.

I will prove facts (1), (4), and (5); you should prove facts (2) and (3).

For (1), let $2x$ and $2y$ denote any two even integers. Then $2x + 2y = 2(x + y)$ [using the distributive property], so the sum has the form $2z$, where $z = x + y$, and thus is even.

In (4), let $2x + 1$ and $2y + 1$ denote any two odd integers. Then
$(2x + 1)(2y + 1) = 4xy + 2x + 2y + 1 = 2(2xy + x + y) + 1$, and hence the product has the form $2z + 1$, where $z = 2xy + x + y$.

Integer* will always mean ***positive integer.*

Finally for (5), any two consecutive integers can be represented by n and $n + 1$ for some integer n. If n is even, i.e. $n = 2q$, then $n + 1 = 2q + 1$ and thus is odd. If n is odd, i.e. $n = 2q + 1$, then $n + 1 = 2q + 1 + 1 = 2q + 2 = 2(q + 1)$, so $n + 1$ is even.

We can also utilize the Division Algorithm for divisors other than 2. For example, if every integer is divided by 3, the possible remainders are 0, 1, or 2; thus every integer can be written in one of the forms $3q$, $3q + 1$, or $3q + 2$. In the same way, every integer can be written in one of the forms $4q$, $4q + 1$, $4q + 2$, or $4q + 3$. These representations can be used to prove:

(6) In every set of three consecutive integers, one integer is a multiple of 3.

(7) In every set of four consecutive integers, one integer is a multiple of 4.

I will prove (6); you should prove (7).

For (6), proceed as you did in (5). Three consecutive integers can be represented as $n, n + 1, n + 2$. If $n = 3q$, then n is a multiple of 3; if $n = 3q + 1$, then $n + 2 = 3q + 1 + 2 = 3q + 3 = 3(q + 1)$, so $n + 2$ is a multiple of 3; and finally if $n = 3q + 2$, then $n + 1$ is a multiple of 3.

I think you now have "enough to go on" to begin Chapter 1. Good luck!

Divide and Conquer

I

Divisibility

In this and the next several chapters, you will investigate the fascinating world of natural numbers (positive integers). It has been said that "God created the integers but all else is the work of man."* Let us begin man's work.

In looking at the structure of the natural numbers or positive integers of our Hindu-Arabic number system (how our number system is "put together"), the most fundamental notion relating two integers is that of "being a factor of."

We say that the integer** a is a factor of the integer b if there exists an integer k such that $b = ka$. We write $a|b$ to denote that a is a factor of b, and write $a \nmid b$ when a is *not* a factor of b. For example,

$3|15$ since $15 = 5 \cdot 3$

and $6|54$ since $54 = 9 \cdot 6$;

however, $3 \nmid 10$ since there is no integer k such that $10 = k \cdot 3$.

For every integer a, $1|a$ and $a|a$, so every integer has at least two factors.

This notion can also be expressed by saying that *a divides b*, or b is a *multiple* of a.

* Leopold Kronecker, a 19th century mathematician.

** "Integer" will always mean *positive* integer unless specified otherwise.

We can graphically represent this concept on a number line; if we locate points for the numbers a and b, the fact that $a|b$ means we can take k copies of the segment from 0 to a and obtain the segment from 0 to b.

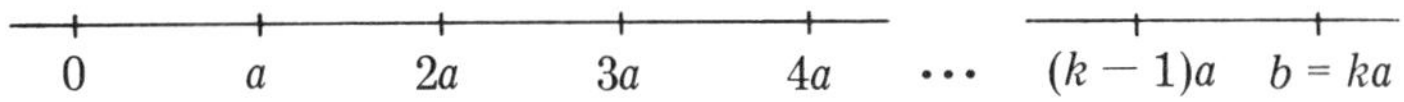

Problem 1. Find all factors of
(a) 10 (b) 24 (c) 32 (d) 210

Problem 2 For this problem consider 0 to be an integer. Which integers are factors of 0?

Problem 3 What is the smallest integer having 4, 6, 9, and 15 as factors?

Problem 4 Give five examples of integers which have exactly
(a) 2 (b) 4 (c) 6 (d) 3 factors.
Do you notice any patterns?

Problem 5 Integer a is said to be a *proper factor* of integer b if $a|b$ and $a < b$. For example, the proper factors of 12 are 1, 2, 3, 4, 6. Find all integers less than 30 which are equal to the sum of their proper factors. Find all integers less than 30 which are equal to the product of their proper factors. (In Chapter 10 we will study in more detail some integers possessing these properties.)

Problem 6 Do there exist any integers x and y such that
(a) $4x + 5y = 77$? (b) $3x + 6y = 77$?
In each case find all pairs of integers (x, y) satisfying the equation or explain why none exist.

Problem 7
(a) How many integers less than or equal to 30 have either 2 or 3 as a factor?
(b) How many integers less than or equal to 900 have either 3 or 5 as a factor?
Explain your reasoning in each case.

Problem 8
(a) How many integers less than or equal to 800 have either 2 or 5, but not 10 as a factor?
(b) How many integers less than or equal to 600 have neither 3 nor 5 as a factor?
Explain your reasoning in each case.

Problem 9
(a) How many integers less than or equal to 60 have either 2, 3, or 5 as a factor?
(b) How many integers less than or equal to 900 have either 2, 3, or 5 as a factor?
Explain your reasoning in each case.

Problem 10 Let $P_3 = \{1\cdot2\cdot3,\ 2\cdot3\cdot4,\ 3\cdot4\cdot5,\ 4\cdot5\cdot6, \ldots\}$. In words, P_3 is the set of all integers which are products of three consecutive integers. What is the largest integer which is a factor of *every* integer in P_3? Explain your reasoning. Now answer the same question for P_4, P_5, P_2, explaining your reasoning in each case. In general, what is the largest integer which is a factor of every integer in P_n?

Problem 11 For what (type of) integers n is $4 \mid (n^2 - n)$? Justify your answer.

Problem 12 You know that $2 \mid 6$, $6 \mid 24$, and $2 \mid 24$. Is this fact always true, i.e. if $a \mid b$ and $b \mid c$, is $a \mid c$? Likewise, $2 \mid 6$, $2 \mid 24$, and $2 \mid (6 + 24)$. Is this fact always true, i.e. if $a \mid b$ and $a \mid c$, is $a \mid (b + c)$? Justify your answer. (Read the text following this problem before attempting it.)

Problem 12 raises the question of mathematical truth. What must be done to answer the question "Is this fact always true?" What is the proper "justification"? To show that the statement "if $a|b$ and $b|c$, then $a|c$" is false, all you need do is find *one* example (usually called a *counterexample*), one instance in which the statement is false. You need to find one set of three integers a_0, b_0, c_0 such that $a_0 \mid b_0$ and $b_0 \mid c_0$, but $a_0 \nmid c_0$.

The same is true for any mathematical statement. For it to be true, it must be true in *every possible case*—not just true some of the time, or most of the time, or even in all but *one* case; there can be no exceptions.

A few examples may help clarify the point. Consider the statements:

(1) Odd integers end in 7.

(2) Odd integers are greater than 10.

(3) Primes (an integer having only itself and 1 as factors) are odd.

All three statements are false; statement (1) might be said to be true some of the time, statement (2) most of the time, and statement (3) in all but one case. But, to reiterate, all three are mathematically false. Give some examples of mathematical statements that are true, and some that are false.

The question of what constitutes a valid proof or justification, or, more importantly, how you discover a proof, has no easy answer. You try to answer the question when you finish this book; I'll try to give you some help along the way.

Problem 13 Determine whether each of the following statements is true or false. For those which are true, construct a proof; for those which are false, find a counterexample.

(a) If $a \nmid b$ and $b \nmid c$, then $a \nmid c$.

(b) If $a \nmid b$ and $a \nmid c$, then $a \nmid (b + c)$.

(c) If $a \mid b$ and $c \mid d$, then $ac \mid bd$.

(d) If $a \mid c$ and $b \mid c$, then $ab \mid c$.

(e) If $a \mid b$, then $a^2 \mid b^2$.

Problem 14 If $a \mid b$ and $b \mid a$, what can you conclude about a and b? Justify your answer.

Problem 15 Consider the first n integers $1, 2, 3, \ldots, n$. Rearrange them in any way you want, listing them as $a_1, a_2, \ldots, a_n$. (For example, 5, 3, 4, 1, 2 is a rearrangement of 1, 2, 3, 4, 5, and in our notation $a_1 = 5$, $a_2 = 3$, $a_3 = 4$, $a_4 = 1$, and $a_5 = 2$.) Form the product $P = (a_1 - 1)(a_2 - 2)\cdots(a_n - n)$. [The product P may be a negative integer or zero.] In the example above, $P = (5 - 1)(3 - 2)(1 - 4)(2 - 5) = 36$.

Show that if n is odd, P is even. (That is, as long as n is odd, no matter how the integers $1, 2, 3, \ldots, n$ are rearranged, the product P is even!) By convention, 0 is considered an even integer.

Much of the value of solving a problem is lost if you do not look back to see what you have done. So try to get into the habit of re-examining your completed solution by asking yourself such questions as:

(1) What was the point to this problem?
(2) What mathematics did I learn from doing this problem?
(3) What did I learn about problem solving by solving this problem?

Look back at the problems in this chapter, especially 6-9, 12, and 13.

HINTS AND COMMENTS

The hints which I give for these problems indicate *one* method of attack—usually there are several. Consequently, you may feel that some of the hints are irrelevant to you, that you can solve the problem in a manner which in no way depends on the hint. Bravo! This shows that you are developing your *own* means of solving problems, which is after all what this little book is trying to get you to do.

Problem 6 What is the essential difference between these two equations?

Problems 7 and 8 Let $n(a)$ denote the number of elements in the set A. Then $n(A \cup B) = n(A) + n(B) - n(A \cap B)$. This statement is sometimes called the "two-set" theorem. Can you make use of this theorem?

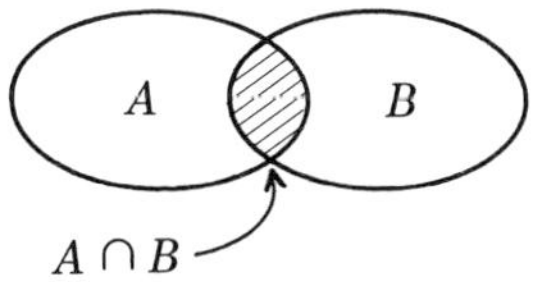

Problem 9 Can you find a "three-set" theorem?

Problem 11 Remember that $n^2 - n = n(n-1)$.

Problem 12 The answer to the first question is true. This problem provides your first example in this book of a proof, so read it carefully.

The expression $a \mid b$ means that there exists an integer k such that $k \cdot a = b$, and $b \mid c$ means there exists an integer ℓ such that $\ell \cdot b = c$. These are the hypotheses. To justify the conclusion $a \mid c$, you must show that there exists an integer, call it m, such that $m \cdot a = c$.

In words, the hypotheses tell you how a is related to b and how b is related to c. How can you use these facts to discover how a is related to c? Symbolically, $k \cdot a = b$, $\ell \cdot b = c$. So, and this is the key, $\ell(ka) = (\ell k)a = c$ by substituting $k \cdot a$ for b in the equation $\ell \cdot b = c$ and using the Associative Law. Thus $m = \ell \cdot k$ is the integer we seek; it is the integer such that $m \cdot a = c$. Thus the statement "if $a \mid b$ and $b \mid c$, then $a \mid c$" is proved. Notice that all these statements are general; they apply to all integers. This is why this argument is a *proof* and not a set of suggestive examples.

You try to prove the other part of the problem.

Problem 13 Reread the proof in Problem 12.

Problem 15 One way to attack this problem is to realize that all that is required for a product of integers to be even is that one of the integers be even. Why, then, are you sure that at least one of the integers $a_1 - 1, a_2 - 2, a_3 - 3, \ldots, a_n - n$ is even?

Note that the conclusion of the problem is false if n is even. For example, rearrange 1, 2, 3, 4 by taking $a_1 = 2, a_2 = 3, a_3 = 4, a_4 = 1$, so that

$$\begin{aligned} P &= (a_1 - 1)(a_2 - 2)(a_3 - 3)(a_4 - 4) \\ &= (2 - 1)(3 - 2)(4 - 3)(1 - 4) \\ &= - 3. \end{aligned}$$

For having lost but once your prime,
You may for ever tarry
— *R. Herrick*

2 Primes

In finding examples of integers having exactly two factors [Problem 4(a), page 10] you undoubtedly obtained a set of integers including 2, 3, 5, 7, 11, 13. Integers having exactly two factors are called *primes,* or stated in a slightly different way, a *prime* is an integer whose only factors are itself and 1. The first twelve primes are

2, 3, 5, 7, 11, 13, 17, 19, 23, 29, 31.

Primes can be thought of as building blocks for the integers in the sense that every integer can be written as the product of primes or is a prime itself. An integer which is not prime is called *composite.* The largest known prime (as of 1971) is $2^{19,937}-1$, an integer having 6,002 digits.

The integer 1 is considered neither a prime nor a composite; 1 is called a *unit.*

After determining a convenient number of small primes, we can use them to test whether larger integers are primes. For example, knowing that 2, 3, 5, 7 are primes, we can conclude that 11 is a prime since $2 \nmid 11$, $3 \nmid 11$, $5 \nmid 11$, $7 \nmid 11$, and that 161 is composite since $2 \nmid 161$, $3 \nmid 161$, $5 \nmid 161$, but $7 \mid 161$. It is not necessary, however, to test all primes less than 11 to show that 11 is a prime. We will consider this idea in Problem 3 below, after you've had a chance to compute a few primes.

It is possible to do some of the following problems by consulting a table of prime numbers. Don't! The purpose of carrying out these computations is to gather evidence upon which you can base your conjectures in later problems.

Problem 1 Find all primes less than 100.

Problem 2 What is the smallest factor greater than 1 of
(a) 20 (b) 35 (c) 221 (d) 191?
Explain carefully why the smallest factor of any composite integer is a prime.

Problem 3 Show that 223 is a prime. What is the largest prime you must test to see whether it is a factor of 223? Why? To decide whether any integer n is a prime, it suffices to check whether all primes less than ____ are factors of n. Explain your reasoning.

Problem 4 Which of the following integers are primes?
(a) 143 (b) 437 (c) 613 (d) 989
(e) 1009.

Problem 5 Find the smallest prime greater than 840.

At this point you should be comfortable with the definition of prime integer and be able (with a little work) to determine whether any "reasonably small" integer is a prime. There is no "easy" way to determine whether a given integer is a prime. If such a method existed, I would not have to say that $2^{19,937} - 1$ is the largest known prime.

And now here are some problems involving primes which should develop your reasoning ability.

Problem 6 Find all primes which are 1 less than a
(a) square (b) cube (c) fourth power.

The use of the word "all" in Problem 6 indicates that some general reasoning is called for. That is, it is impossible to look at *every* prime individually and decide whether or not it is 1 less than a square. Thus your solution to this problem will be something like:

"The primes $3, p_2, p_3, \ldots, p_k$ are all 1 less than a square because I numerically verified it, e.g. $3 = 2^2 - 1$, and all other primes cannot be 1 less than a square because...."

Problem 7 Which integers in the set $\{7, 17, 27, 37, 47, \ldots\}$ can be written as the sum of two primes? Explain your reasoning.

Problem 8 Which primes cannot be written as the sum of two composite integers? Justify your answer. (Remember that 1 is not considered a composite.)

Problem 9 Which primes (if any) can be written as the sum of
(a) 2 (b) 3 (c) 4 (d) 6 consecutive integers?
Explain your reasoning in each case.

Problem 10 An *arithmetic progression* (of integers) is a sequence of integers of the form $a,\ a + b,\ a + 2b,\ a + 3b, \ldots,$ where a and b are integers. For example, $3, 7, 11, 15, 19, \ldots$ is an arithmetic progression with $a = 3$ and $b = 4$. The integers $3, 7, 11, 15, 19$, etc. are called the *terms* of the progression.

(a) Write down the first six terms of the arithmetic progression where
(i) $a = 2,\ b = 3$ (ii) $a = 5,\ b = 8$ (iii) $a = 2,\ b = 4$.

(b) Can every term of an arbitrary arithmetic progression be a prime, i.e. for any two integers a and b, is it possible that $a,\ a + b,\ a + 2b,\ a + 3b, \ldots$ are all primes? Justify your answer.

(c) Does every arithmetic progression have a term which is a prime? Make a conjecture on how you can tell whether an arithmetic progression has a term which is a prime.

Problem 11 Let $E = \{2, 4, 6, 8, 10, \ldots\}$. Note that E is closed with respect to multiplication, i.e. the product of any two elements in E is again an element of E since $(2k)(2\ell) = 2(2kl)$. A "*pryme*" in E is an integer n in E which cannot be expressed as the product of two integers *in E*, both of which are less than n. For example, $6, 10, 14$ are "prymes," while $8 = 2 \cdot 4$ and $24 = 6 \cdot 4$ are "composites." Which elements of E are "prymes" in E?

Problem 12 Let $T = \{3, 6, 9, 12, 15, 18, \ldots\}$. Is T closed with respect to multiplication? Why? Which elements of T are "prymes" in T?

Problem 13 Let $M = \{5, 9, 13, 17, 21, 25, 29, 33, \ldots\}$; M consists of all integers of the form $4n + 1$. Show that M is closed with respect to multiplication. Find four "prymes" and four "composites" in M. ("Pryme" and "composite" are defined as in Problem 11.)

The final problem in this chapter illustrates another important method of reasoning—*indirect reasoning.* Briefly, the process of indirect reasoning consists of (1) determining all possible alternatives to a given situation (usually there are only two—either something occurs or it doesn't occur) including the one you hope to be true, and then (2) proving that by assuming the truth of any alternative other than the one you are trying to prove true, you can logically deduce a false statement. Thus the truth of your assertion is established by a process of elimination of the other alternatives.

Problem 14 Prove that every integer of the form $4k + 3$ has a prime factor of that form. For example, $35 = (4\cdot\underline{8}) + 3$ and $35 = 5\cdot 7$. Now, 5 is not of the form $4k + 3$, but 7 is: $7 = (4\cdot\underline{1}) + 3$. (What are the possible alternatives? Which one do you wish to prove? How can you deduce a false statement by assuming the truth of any other alternative?)

HINTS AND COMMENTS

Problem 3 In determining the largest prime which need be tested to show that 223 is prime, you might consider the following idea. If a prime p were a factor of 223, how large would the other factor, $223/p$, be? Ask this question for each prime you test. That is, if you express an integer as the product of two factors, as one factor gets larger, the other gets smaller.

Problem 6 Another fundamental principle of problem solving is illustrated for the first time in this problem—the principle of translating the problem into mathematical language. In our context, this means setting up an algebraic equation which represents the statement of the problem. In Problem 6, we wish to "find all primes which are 1 less than a square." As an algebraic equation, then, the problem becomes one of finding all values of p for which $p = n^2 - 1$ for some integer n. But $p = n^2 - 1 = (n - 1)(n + 1)$, so you must now decide: for what values of n can the product of the two integers $(n - 1)$ and $(n + 1)$ be a prime?

Problem 8 Here is a problem which illustrates a general method of attacking problem solving—if such a thing even exists. First try some examples:

2 and 3 cannot be written as the sum of two composite integers

$5 = 2 + 3$ so 5 " " "

$7 = 2 + 5$ so 7 " " "
$3 + 4$

$11 = 2 + 9$ so 11 " " "
$3 + 8$
$4 + 7$
$5 + 6$

$13 = 2 + 11$ so 13 *can* be written as the sum of two composite integers
$3 + 10$
$4 + 9$

$17 = 8 + 9$ so 17 " " "

$19 = 9 + 10$ so 19 " " "

On the basis of these examples, you are probably led to the conjecture that 2, 3, 5, 7, and 11 are the only primes which cannot be written as the sum of two composites.

Check this conjecture in a few more examples!

$23 = 9 + 14$ so . . . ?
$\quad\; 8 + 15$

$29 = 9 + 20$ so . . . ?
$\quad\; 4 + 25$

So now the problem is to decide: Why can *any* prime p greater than 11 be written as the sum of two composite numbers? Look back at the examples and see whether you notice any patterns among the primes 13, 17, 19, 23, 29. Then ask yourself the key question: "Is this pattern which is true for the primes 13, 17, 19, 23, 29 true for any prime?" Thus if you can find a general reason why *any* prime greater than 11 can be written as the sum of two composite numbers, you have completely solved this problem. Find it!

Problem 9 Write an algebraic equation which says that the prime p is the sum of 2, 3, 4, or 6 consecutive integers; i.e. the key question then is: how can you algebraically represent a sum of 2, 3, 4, or 6 consecutive integers?

Problem 10 After trying several examples, you have probably conjectured that every arithmetic progression has at least one composite term. Thus here again you must find a general reason, i.e. why is it true that *no matter what* pair of integers a and b are chosen, at least one of the integers $a, a + b, a + 2b, a + 3b,\ldots$ must be composite. A list of examples will not suffice.

Problem 14 As I mentioned, this problem is your first example of the use of indirect reasoning. You wish to prove that every integer of the form $4k + 3$ has a prime factor of that same form.

Let n denote any integer of the form $4k + 3$, i.e. $n \in \{3, 7, 11, 15, 19, 23,\ldots\}$. Then you wish to prove at least one prime factor of n is also in this set. Thus there are two alternatives:

(1) At least one prime factor of n belongs to $\{3, 7, 11, 15, \ldots\}$.

(2) No prime factor of n belongs to the set.

One of these two statements must be true. To prove that our desired conclusion, statement (1), is true, we will assume that statement (2) is true and derive a contradiction to a known fact. This contradiction will mean, then, that statement (2) cannot be true; hence statement (1) must be true.

Let us suppose, now, that statement (2) is true; that is, suppose no prime factor of n has the form $4\ell + 3$. This means that every prime factor of n has the form $4\ell + 1$ since every prime (except 2) is either of the form $4\ell + 1$ or $4\ell + 3$. (Why?) But if every prime factor of n is of the form $4\ell + 1$, then n also has that form since $n = (4\ell_1 + 1)(4\ell_2 + 1) \cdots (4\ell_n + 1) = 4L + 1$ for some integer L; or, stated another way, the set M of Problem 13 is closed. Hence we have contradicted the fact that n has the form $4k + 3$. Thus the supposition that n has no factor of the form $4\ell + 3$ led to the conclusion that n has the form $4k + 1$, so the supposition is false. Therefore, (2) cannot be true, leaving only the possibility that (1) is true, i.e. that indeed n has a prime factor of the form $4\ell + 3$.

Reread this proof until you understand it thoroughly.

(After) Problem 14 It is especially critical that you carefully review what you have done in this chapter. In addition to the mathematics you've learned, you've been exposed to two basic techniques of problem solving: (1) expressing a problem in mathematical language, and (2) indirect reasoning.

3

For right timing is in all things
the most important factor.

Factorization

The fact that the set of primes forms the building blocks for the set of integers is one of the most fundamental concepts of elementary arithmetic. (For example, the primes 2, 2, 3, 5 can be multiplied together to form the integer 60, i.e. $2 \cdot 2 \cdot 3 \cdot 5 = 60$.) This idea is so important that it is usually referred to as the

Fundamental Theorem of Arithmetic

Every integer n greater than 1 is expressible uniquely as a product of primes; in symbols,

$$n = q_1^{a_1} q_2^{a_2} \cdots q_k^{a_k}$$

where the factors q_i are primes and the exponents a_i represent the number of times a prime q_i occurs as a factor of n.

For example,

$$210 = 2 \cdot 3 \cdot 5 \cdot 7$$

$$4725 = 3^3 \cdot 5^2 \cdot 7 \qquad (\text{Here } q_1 = 3,\ a_1 = 3;\ q_2 = 5,\ a_2 = 2;\ q_3 = 7,\ a_3 = 1)$$

We regard $2 \cdot 3 \cdot 5 \cdot 7$, $3 \cdot 2 \cdot 7 \cdot 5$, $7 \cdot 3 \cdot 5 \cdot 2$, etc. as the same factorization; the order in which the prime factors are considered does not matter. The Fundamental Theorem, then, means that for every integer n there is only one set of primes which when multiplied together yields n.

Assuming the Fundamental Theorem to be true, we will call $2 \cdot 3 \cdot 5 \cdot 7$ the *prime factorization* of 210; similarly, $3^3 \cdot 5^2 \cdot 7$ is the prime factorization of 4725.

Problem 1 Find the prime factorizations of
(a) 9240 (b) 5445 (c) 3599 (d) 158,000.

The Fundamental Theorem in essence means that Problem 1 has a solution. The sentence "find the prime factorization of n" implies two things: (1) that n *has* a prime factorization, i.e. a prime factorization of n exists, and (2) there is only one prime factorization of n (because of the sense of the word "the"). These two things are exactly what is guaranteed by the Fundamental Theorem.

An analogy which may be meaningful to some of you is the one between integers and molecules. If one thinks of atoms as primes and molecules as composites, then the fundamental notion that all molecules are made up of certain atoms arranged in a unique way is analogous to the Fundamental Theorem. For example, compare the following facts:

(1) A sulfuric acid molecule, H_2SO_4, is made up of two hydrogen atoms, one sulfur atom, and four oxygen atoms.

(2) $7500 = 2^2 \cdot 3 \cdot 5^4$.

Problem 2 For what (type of) integers n is the product of all proper factors of n less than n? Explain your reasoning.

Problem 3 Find the prime factorization of 24,949,501 given that its smallest prime factor is 499.

Problem 4 Find the smallest (a) square (b) cube which has 180 as a factor. Explain your reasoning. Do the same using 1440 as a factor.

Problem 5 Determine whether each of the following statements is true or false. Justify your answers.

(a) Suppose that $k^2 = 3ab$ for some integers k, a, b.
Then (i) $3 \mid k$ (ii) $3 \mid a$ or $3 \mid b$
(iii) $3 \mid a$ and $3 \mid b$.

(b) Suppose that $n^2 = 12ab$ for some integers n, a, b.
Then (i) $12 \mid n$ (ii) $12 \mid a$ or $12 \mid b$
(iii) $12 \mid a$ and $12 \mid b$.

Problem 6 What is the smallest integer greater than 1 which is simultaneously

(a) a square and a cube?

(b) * a square, a cube, and a fifth power?

In general, if the prime factorization of $n = q_1^{a_1} q_2^{a_2} \cdots q_k^{a_k}$, what can you say about the exponents a_i if n is

(c) a square (d) a cube (e) a fifth power

(f) an mth power?

Applications of the Fundamental Theorem are many, as evidenced by the remaining problems in this chapter. Students have found these problems more difficult than those in the first two chapters. The reason is that the Fundamental Theorem is deceptive; its meaning is clear but its proof is "beyond the scope of this book" (the customary remark in such instances). Some of these problems are my attempts to focus on some of the ideas involved in this proof. Nonetheless, spend enough time on these problems; it will be worth your while because you will then really understand the Fundamental Theorem and its consequences. These problems should also convince you that many interesting problems can be constructed which depend on a seemingly simple idea.

* It is sufficient to find the prime factorization (using exponential notation) for this integer.

Problem 7 Can the sum of two different primes ever be a factor of the product of those two primes? i.e. do there exist primes p and q such that $(p + q) \mid pq$? Either find an example or explain why none exists.

Problem 8 We define $k! = k(k - 1) \cdots 3 \cdot 2 \cdot 1$ or, in words, k *factorial* is the product of all integers less than or equal to k. Thus $3! = 3 \cdot 2 \cdot 1 = 6$, $4! = 4 \cdot 3 \cdot 2 \cdot 1 = 24$, $5! = 5 \cdot 4 \cdot 3 \cdot 2 \cdot 1 = 120$.

(a) For what (type of) integers n is $n \mid (n - 1)!$? Justify.

(b) Suppose we leave out every other integer in $(n - 1)!$ and ask the same question, i.e. for what (type of) integers n is $n \mid (n - 1)(n - 3)(n - 5) \cdots 2$ (or 1)? (The final integer is 2 if n is odd, and 1 if n is odd, and 1 if n is even.) Justify.

Problem 9 There are no zeros at the end of $3! = 6$, one zero at the end of $5! = 120$, and two zeros at the end of $10! = 3{,}628{,}800$. How many zeros are there at the end of $100!$? (There are more than 20.) Explain your reasoning. How many zeros are there at the end of $1000!$?

Problem 10 Compute $(n - 1)! + 1$ for $n = 2, 3, 4, 5, 6, 7, 8, 9$. For which of these integers is $n \mid (n - 1)! + 1$? In general, for what (type of) integers would you conjecture that $n \mid (n - 1)! + 1$?

Problem 11

(a) Find all primes p such that the integer $5p + 1$ is a square. Justify your answer.

(b) Find all primes p for which $7p + 4$ is a square. Justify your answer.

Problem 12 Determine whether each of the following statements is true or false. Justify your answers.

(a) If $a^2 \mid b^2$, then $a \mid b$.
(b) If $a^a \mid b^b$, then $a \mid b$.

Problem 13 It is clear that if $6 \mid n$ for some integer n, then $6 \mid n^2$. But suppose we interchange the hypothesis

(a) if $6 \mid n^2$, is $6 \mid n$?
(b) if $12 \mid n^2$, is $12 \mid n$?
(c) In general, for what (type of) integers m is the following statement true: If $m \mid n^2$, then $m \mid n$.

Explain your reasoning if you can, especially in **(a)** and **(b)**.

Problem 14 Find the least integer which is:

(a) one-half of a square *and* one-third of a cube.
(b) one-half of a square, one-third of a cube, *and* one-fifth of a fifth power.*
(c) one-half of a square, one-third of a cube, one-fifth of a fifth power, *and* one-seventh of a seventh power.*

* It is sufficient to find the prime factorization (using exponential notation) for these integers.

Problem 15

(a) Do there exist integers x and y such that

(i) $x^2 - y^2 = 48$?

(ii) $x^2 - y^2 = 23$?

(iii) $x^2 - y^2 = 45$?

(iv) $x^2 - y^2 = 90$?

In each case find all integers x, y which satisfy the equation, or explain why none exist.

(b) In general, for what (type of) integers n does $x^2 - y^2 = n$ have a solution?

Problem 16 The Fundamental Theorem, in essence, states that "prime factorizations are unique in the set of (positive) integers". This fact is not, however, true for all sets of integers. For example, in the set $E = \{2, 4, 6, 8, 10, 12,...\}$ (Chap. 2, page 20) we saw that 2, 6, and 18 were "prymes". Now, $36 = 6 \cdot 6 = 2 \cdot 18$; thus we have expressed 36 as a product, *in E*, of "prymes" in two *different* ways, so that prime factorizations, *in E*, are not unique.

(a) Find three other numbers in E having two different prime factorizations.

(b) Show that for the sets $T = \{3, 6, 9, 12, 15,...\}$ (Chap. 2, page 20) and $M = \{5, 9, 13, 17, 21, 25,...\}$ (Chap. 2, page 21) prime factorizations are not unique by finding at least two integers in each set having two different prime factorizations.

HINTS AND COMMENTS

Problem 3 See Problem 3, Chapter 2 (page 18).

For the remaining problems, try to think of an integer in terms of its prime factorization.

Problems 4 and 6 Give the prime factorizations of the desired integer as your answer. You will then see "what's going on" in these problems.

Problem 8 In testing examples, it is not necessary to compute $(n-1)!$ each time. For example, in testing whether $12 \mid 11!$, write $2 \cdot 6 \mid 11 \cdot 10 \cdot 9 \cdot 8 \cdot 7 \cdot \underline{6} \cdot 5 \cdot 4 \cdot 3 \cdot \underline{2}$, so $12 \mid 11!$. That is, don't multiply out $(n-1)!$ each time, it gets much too large! Instead, write n as the product of two factors, and see whether both of those factors occur in $(n-1)!$.

Problem 11 Again you can use your high school algebra here (as well as the Fundamental Theorem). In algebraic language, the problem asks you to find all primes p such that the equation $5p + 1 = n^2$ has a solution for some integer n. But if $5p + 1 = n^2$, then $5p = n^2 - 1 = (n+1)(n-1)$.

Now, the Fundamental Theorem says every integer can be written *uniquely* as a product of primes. Here we have an integer, $5p$, expressed as a product of primes. However, we also have expressed it another way, as $(n-1)(n+1)$. But there is only *one* prime factorization, so we must have that

$$\begin{array}{ccc} n-1=5 & & n+1=5 \\ & \text{or} & \\ n+1=p & & n-1=p. \end{array}$$

Be sure this last statement is clear to you.

Problem 13 See Problem 6, Chapter 3 (page 27).

Problem 14(a) Let n denote the least integer, then $n = x^2/2$ for some integer x, and $n = y^3/3$ for some integer y. Hence $2 | n$ and $3 | n$. (Why?) So $n = 2^a 3^b$ and the problem is to determine the exponents a and b. Since $n = x^2/2$, $2n = x^2$; thus $2(2^a 3^b) = 2^{a+1} 3^b$ is a square, and hence $a + 1$ and b are even. (Why? See Problem 6 above.) Similarly, $n = y^3/3$ so $3n = y^3$; thus $3(2^a 3^b) = 2^a 3^{b+1}$ is a cube, and hence a and $b + 1$ are multiples of 3. Thus a is the smallest integer such that $a + 1$ is even and a is a multiple of 3, i.e. $a = 3$. Similarly, $b = 2$, so $n = 2^3 3^2 = 72$. Be sure you understand the reasoning here before you go on to parts **(b)** and **(c)**.

Problem 15 $x^2 - y^2 = 48$. Now, $x^2 - y^2 = (x - y)(x + y)$, so how many ways can 48 be expressed as the product of two integers? $48 = 1 \cdot 48$, $48 = 2 \cdot 24$, $48 = 3 \cdot 16$, $48 = 4 \cdot 12$, $48 = 6 \cdot 8$. For which of these factorizations can $x + y$ represent the larger factor and $x - y$ represent the smaller one? For example, do there exist integers x, y such that

$$\begin{aligned} x + y &= 48 \\ x - y &= 1 \end{aligned} \qquad \text{or} \qquad \begin{aligned} x + y &= 24 \\ x - y &= 2, \end{aligned} \qquad \text{etc. ?}$$

REFLECTIONS ON PROBLEM SOLVING

I

If you're new to the experience of problem solving, (as I suspect many of you are), you're probably somewhat unsure of yourself at this point. You ask yourself: "How do you solve a problem?" "How do you justify your answer?" "How do you explain your reasoning?" Don't despair! This uncertainty is actually a good sign, for you're beginning to ask questions about concepts you never questioned before. You're trying to probe into the "whys" rather than just the "hows."

Let's now attempt to partially answer these questions by analyzing some of the key ideas about problem solving which were brought out in the first three chapters.

(1) First try several numerical examples. This should help you see what the problem is all about, help you reinterpret the problem in a way meaningful to you.

(2) Then make a conjecture based on these examples.

(3) Check this conjecture with a few more examples.

(4) Verify the conjecture.

Step (4) is certainly the most difficult. We have seen two general methods to go about it. The first could be termed *direct reasoning* (e.g. Chap. 1, Prob. 12). The reasoning takes the form of a chain of statements: if A is true, then B is true; if B is true, then C is true; and so on until the conclusion is deduced. The second, *indirect reasoning* (e.g. Chap. 2, Prob. 14), showing that any conclusion other than the desired one is false, and therefore the desired conclusion must be true by a process of elimination. Two specific suggestions were also made:

(A) Translate the problem into mathematical language; set up an algebraic equation which represents the statement of the problem.

(B) Look back and assess what you did.

While the foregoing discussion is necessarily vague, it must be realized that there is no (at least the author has never found one) specific method which will yield a solution to any given problem. If such a procedure existed, there would certainly be a lot fewer problems in the world today. The ability to solve problems is largely acquired through experience. Hopefully, when you are confronted by a problem, you will now do more than just sit and look at it. You may not always succeed, but in making an effort you will be thinking, probing, analyzing, experimenting, creating; and certainly mental activities such as these have some value, for "what you have been obliged to discover by yourself leaves a path in your mind which you can use again when the need arises" (to paraphrase from the German philosopher Lichtenberg).

Let your fingers do the walking
— Bell Telephone Company

4 Digits

Now you (with a little help) are going to develop some tools for finding the prime factorizations of integers. One such tool is to analyze the relationship between the digits of an integer and some of its factos; consequently we will now discuss digits. At the same time we will take a close look at what the place-value concept really means.

One of the axioms of arithmetic which is the basis for the place-value approach to our Hindu-Arabic number system is sometimes referred to as the

Principle of Expanded Form (base 10)

Every integer N can be written uniquely in the form

$$N = a_n \cdot 10^n + a_{n-1} \cdot 10^{n-1} + \cdots + a_1 \cdot 10 + a_0$$

where the a_i are integers with $0 \leq a_i \leq 9$ for $i = 0, 1, 2, 3, \ldots, n$. The a_i are the *digits* of N.

For example, $24{,}376 = 2 \cdot 10^4 + 4 \cdot 10^3 + 3 \cdot 10^2 + 7 \cdot 10 + 6$. (Here $n = 4$, $a_0 = 6$, $a_1 = 7$, $a_2 = 3$, $a_3 = 4$, $a_4 = 2$.) The number 6 here is called the *unit's digit,* 7 the *ten's digit,* 3 the *hundred's digit,* 4 the *thousand's digit* and 2 the *ten thousand's digit.*

This representation is extremely useful for solving many types of problems as well as being the principle upon which many arithmetical oddities and recreational puzzles are based.

Problem 1 It takes 867 digits to number the pages of a book. (Page 9 uses one digit, page 37 uses two digits, etc.)
(a) How many pages are there in the book?
(b) How many times does the digit 5 appear?

In the next set of problems, you can make extensive use of the principle of translating a problem into mathematical language by setting up an algebraic equation which represents the problem. You will constantly use the fact that any two-digit integer can be algebraically represented by the quantity $10a + b$; any three-digit integer by $100a + 10b + c$, etc.

Problem 2
(a) A 2-digit integer is added to the integer formed by reversing its digits. This sum always has the integer ___ as a factor. Justify your answer.
(b) Take any 3-digit integer (e.g. 638), reverse the digits, and find the non-negative difference between these two integers (e.g. $836 - 638 = 198$). This difference always has the integer ___ as a factor. Justify your answer.

Problem 3
(a) Show that the product of the digits of any 2-digit integer is always less than that integer (e.g. $5 \cdot 7 = 35 < 57$).
(b) Find all 2-digit integers which equal *twice* the product of their digits. Use your head instead of your pencil.
(c) Formulate and solve another problem relating a 2-digit integer and the sum of its digits.

Problem 4 A *palindrome* is an integer (or a word, phrase, etc.) which reads the same forwards as backwards. For example, 131, 2552, 34,743 are palindromes.

(a) How many 2-digit palindromes are there? 3-digit?

(b) What integer greater than 1 is a factor of every 4-digit palindrome? Prove your answer.

Problem 5 Find all 2-digit integers such that the product of that integer and the integer formed by reversing its digits is a 3-digit palindrome.

Problem 6 Formulate and solve a problem involving palindromes.

And now a look at some of the properties of squares which depend on their digits.

Problem 7 Guess the rule for squaring (without peforming any computations on paper) suggested by the following examples: $25^2 = 625$, $45^2 = 2025$, $65^2 = 4225$. Compute 35^2, 55^2, 75^2, 85^2, 95^2, using your rule. Justify your rule.

Problem 8 What integers can occur as the unit's digit of a square?

Problem 9 Find all integers n for which the integer $(1! + 2! + 3! + \cdots + n!)$ is a square.

The next three problems will enable you to look closely at what *place value* really means.

Problem 10 Which integers in the set $\{1, 11, 111, 1111, 11111, \ldots\}$ are squares? Why? Or, for that matter, does there exist a square (of two or more digits), all of whose digits are the same? Justify.

Problem 11 Find all 2-digit integers whose squares have
(a) four even digits,
(b) four odd digits.
Describe any time-saving procedures you used or any general method of attacking this problem.

The next problem is a good one to test your ability to recognize patterns.

Problem 12 Examine the following numerical relationships:

$$12 \cdot 42 = 21 \cdot 24$$
$$24 \cdot 63 = 42 \cdot 36$$
$$46 \cdot 96 = 64 \cdot 69.$$

Find three more examples of this relationship. Find and justify a rule (depending on the digits of the integers involved) for determining all pairs of 2-digit integers related in this way. (Omit such examples as $57 \cdot 75 = 75 \cdot 57$.)

We close this chapter with three problems of a more recreational nature. The final problem can certainly be classified as an arithmetical oddity. Such oddities are part of the fascination of numbers which have intrigued its students for centuries.

Problem 13 Using each of the nine digits (not 0) only once, find primes which have the smallest possible sum. For example, $61 + 283 + 47 + 59 = 450$, but there is a smaller sum which can be formed in an analogous way. Find it.

Problem 14 There are many sets of integers which can be formed by using each of the ten digits only once. For example, $\{347, 256, 9, 810\}$, $\{12, 34, 56, 78, 90\}$, $\{1, 23, 456, 7890\}$, etc. Can you form such a set where the sum of the integers in the set is 100? Either find an example or explain why none exists.

Problem 15 Take any 3-digit integer (e.g. 638), reverse the digits, and find the non-negative difference between these integers (e.g. $836 - 638 = 198$). Now reverse the digits of this difference, and add these two integers ($198 + 891 = 1089$). If the difference is a 2-digit integer, insert a 0 in the hundred's place. What integers can occur as the final sum (e.g. 1089)? (Try some examples.) Justify your answer if you can.

HINTS AND COMMENTS

The best hint I can give you for these problems is to represent the problem algebraically and then do exactly what the problem says. In most cases an equation or algebraic expression will evolve, and a close examination of that equation or expression should yield a solution. For example, in Problem 2, $(10a + b) + (10b + a) = 11(a + b)$.

Problem 10 Here's a real chance to look at what it means to multiply two integers. If the square of an integer n has unit's digit 1, what can be the unit's digit of the integer? Answer: 1 or 9. If the unit's digit of n is 1, what can you say about the ten's digit of its square? Schematically the problem looks like

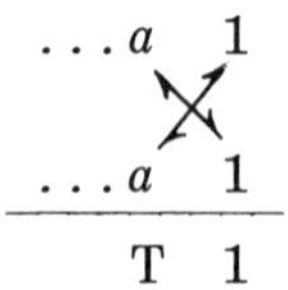

The ten's digit T is the remainder when $a + a$ is divided by 10. Why? Can this digit be 1?

If the unit's digit of n is 9, schematically we have

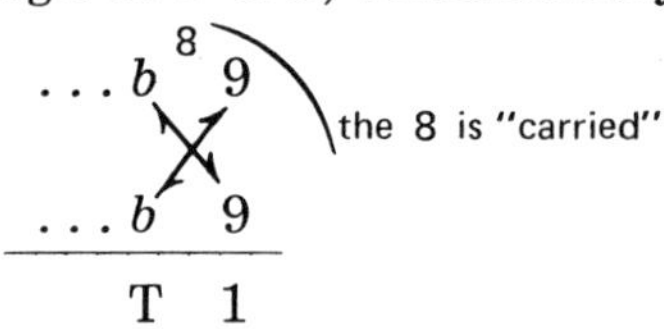

and the ten's digit T of the square is the remainder when $9b + 9b + 8 = 18b + 8$ is divided by 10. Why? Can this digit be 1?

In doing the problem for digits other than 1, consider the case $444\cdots4$ last.

5 Divisibility Tests

United we stand, divided we fall.

In this chapter, as we said in Chapter 4, we are going to try to relate the factors of an integer to its digits; that is, knowing the digits of an integer (which we do, of course, if we know the integer), can we determine any factors of the integer? For example, we know by inspection that 10 is a factor of 12,780 because the unit's digit of 12,780 is 0. This example illustrates the use of the two most obvious divisibility tests:

Divisibility by 2 An integer has a factor of 2 if its unit's digit has a factor of 2.

Divisibility by 5 An integer has a factor of 5 if its unit's digit is either 0 or 5.

Problem 1 Explain carefully why these two tests work.

The divisibility test for 3 makes use of all the digits of the integer. It is, perhaps, the best example of relating the digits of an integer to one of its factors.

Divisibility by 3 An integer has a factor of 3 if the sum of its digits has a factor of 3.

For example, $3|234{,}756$ since $3|(6+5+7+4+3+2=27)$.

Problem 2 Give three more examples illustrating this criterion. Prove the criterion.

Problem 3 Devise a criterion for divisibility by 9. Give some examples. Give two examples (each having at least six digits) of integers having 3, but not 9, as a factor. Finally prove your criterion.

Problem 4 Devise a criterion for divisibility by 4, and by 8. Give several examples illustrating the use of each of your criteria. Prove each of your criteria.

You have seen a special case of the following test when you examined 4-digit palindromes (Problem 4(b), Chapter 4, page 37).

Divisibility by 11 The integer

$$N = a_n 10^n + a_{n-1} 10^{n-1} + \cdots + a_1 10 + a_0$$

has a factor of 11 if the "alternating sum of its digits," that is, if the integer $a_0 - a_1 + a_2 - a_3 + \cdots \pm a_n$ has a factor of 11.

Examples:

(a) $11 \mid 182{,}347$ since $11 \mid (7 - 4 + 3 - 2 + 8 - 1 = 11)$

(b) $11 \mid 87{,}564{,}312$ since $11 \mid (2 - 1 + 3 - 4 + 6 - 5 + 7 - 8 = 0)$

because every integer is a factor of 0.

Problem 5 Give a few more examples illustrating this criterion. Prove the criterion.

The next problem is a "divisibility by 7" test for 3-digit integers which can be used as a basis for a general "divisibility by 7" test coming up in Problem 7.

Problem 6 Let n be a 3-digit integer, i.e. let $n = 100a + 10b + c$.

Criterion: If $2a + 3b + c$ has a factor of 7, then n has a factor of 7.

(a) Illustrate this criterion with several examples. Then prove the criterion.

(b) Devise and prove a similar criterion for 4-digit integers.

Problem 7 Examine the following examples:

(i) Both 1,516,592 and $592 - 516 + 1 = 77$ have 7 as a factor.

(ii) Both 89,287,653 and $653 - 287 + 89 = 455$ have 7 as a factor.

(iii) Both 10,458,742 and $742 - 458 + 10 = 294$ have 7 as a factor.

Devise a criterion based on these examples for divisibility by 7. (Note that the examples show that you can tell whether any integer is divisible by 7 by examining only a 3-digit integer). Give some examples illustrating your criterion, and finally (of course) prove your criterion.

Problem 8 Show that the criterion you developed in Problem 7 is also valid for divisibility by 11, and by 13. Give some examples first.

Problem 9 Using the above divisibility tests as an aid, find the prime factorizations of the integers

(a) 207 (b) 1065 (c) 708 (d) 1080
(e) 4807 (f) 54,747 (g) 3,602,599.

Problem 10 Is the integer 42,571,362,444 a square? Explain your reasoning.

Problem 11 Find all possible values for the digits a and b such that (i) $36 \mid a{,}89b$ and (ii) $99 \mid 3a4{,}9b5$. Save some time and use your head instead of your pencil.

Problem 12 Which integers in the set
$\{9, 98, 987, 9876, ..., 987654321, 9876543219, 98765432198, ...\}$
are composite? Explain your reasoning.

Problem 13

(a) Multiplication by (i) 4 (ii) 9 reverses the digits of some 4-digit integers. Find all such integers in each case, and explain your reasoning.

(b) Is this phenomenon possible for any 1-digit integer other than 4 or 9? Why?

Thus far we have studied all our arithmetical concepts with respect to base 10. However, corresponding notions also exist in all other bases. As you probably know, many facts which are true about integers in base 10 are not true in some other base, and conversely. Let's investigate the situation with regard to the divisibility tests. Hopefully, examining these tests in another base will reinforce your understanding of them in base 10. (I assume that you have some familiarity with expressing an integer n in base b, and with performing the four basic arithmetical operations in it.)

The Principle of Expanded Form is valid for any base b, i.e. every integer $N_{(b)}$ can be written in the form

$$N_{(b)} = a_n \cdot b^n + a_{n-1} \cdot b^{n-1} + \cdots + a_1 \cdot b + a_0$$

where the a_i are integers with $0 \le a_i \le b-1$ for $i = 0, 1, 2, 3, ..., n$. For example,

$$2341_{(7)} = (2 \cdot 7^3) + (3 \cdot 7^2) + (4 \cdot 7) + 1$$

and

$$101101_{(2)} = (1 \cdot 2^5) + (1 \cdot 2^3) + (1 \cdot 2^2) + 1.$$

Problem 14 Is the divisibility test for 3 valid in base 7? in base 5? For what bases b is it valid? Justify your answer if you can.

Problem 15 For what bases b is your divisibility-by-9 test valid? Justify your answer if you can.

Problem 16 For what bases b is your divisibility-by-4 test valid? Justify your answer if you can.

Problem 17 For those bases b for which any of the above divisibility tests were not valid, see if you can formulate a new divisibility test. For example, find
(a) a "divisibility by 4" test for base 5,
(b) a "divisibility by 5" test for base 9,
(c) a "divisibility by 9" test for base 12.

HINTS AND COMMENTS

Problem 2 Let's first look at a specific integer which we know has a factor of 3, and analyze why this divisibility test works for that integer.

Example. $3 \mid 7{,}482$ and we observe that $3 \mid (2 + 8 + 4 + 7 = 21)$; but why does this imply that $3 \mid 7{,}482$? Now, $7{,}482 = D + (2 + 8 + 4 + 7)$ for some integer D. If we could show that $3 \mid D$, then we could conclude that $3 \mid 7{,}482$ since $3 \mid (2 + 8 + 4 + 7)$. We are using the fact that if $a \mid b$ and $a \mid c$, then $a \mid (b + c)$ where in this case $a = 3$, $b = D$, and $c = (2 + 8 + 4 + 7)$. (See page 12.) So we now have to decide why $3 \mid D$. What is D?

We know that $7{,}482 = 7 \cdot 10^3 + 4 \cdot 10^2 + 8 \cdot 10 + 2$,

so $$D = 7 \cdot 10^3 + 4 \cdot 10^2 + 8 \cdot 10 + 2 - (2 + 8 + 4 + 7)$$

$$= 7(10^3 - 1) + 4(10^2 - 1) + 8(10 - 1)$$

$$= 7 \cdot 999 + 4 \cdot 99 + 8 \cdot 9$$

so that certainly $3 \mid D$ since $3 \mid 9$, $3 \mid 99$, and $3 \mid 999$.

Now let's try to prove the criterion for any 4-digit integer, call it $N = a_3 \cdot 10^3 + a_2 \cdot 10^2 + a_1 \cdot 10 + a_0$. Let S denote the sum of the digits, i.e. $S = a_0 + a_1 + a_2 + a_3$, and suppose (as the criterion says) that $3 \mid S$. We see that $3 \mid (N - S)$ since

$$N - S = a_3(10^3 - 1) + a_2(10^2 - 1) + a_1(10)$$
$$= (a_3 \cdot 999) + (a_2 \cdot 99) + (a_1 \cdot 9).$$

Now, $N = (N - S) + S$, so since $3 \mid S$ and $3 \mid (N - S)$, we conclude that $3 \mid N$.

Review this proof carefully. Nowhere did we use the fact that the integer N had 4 digits. This means that the same proof should be valid no matter how many digits N has; that is, we should be able to write a general proof which is valid for any integer N. Let's do just that.

General Proof

Suppose $N = a_n 10^n + a_{n-1} 10^{n-1} + \cdots + a_1 10 + a_0$. Let S denote the sum of the digits of N, i.e. $S = a_0 + a_1 + a_2 + \cdots + a_n$, and suppose $3 \mid S$. Now, $3 \mid (N - S)$ since

$$N - S = a_n(10^n - 1) + a_{n-1}(10^{n-1} - 1) + \cdots + a_2(10^2 - 1) + a_1(10 - 1)$$
$$= (a_n \cdot \underbrace{999\cdots9}_{n \text{ times}}) + (a_{n-1} \cdot \underbrace{999\cdots9}_{(n-1) \text{ times}}) + \cdots + (a_2 \cdot 99) + (a_1 \cdot 9).$$

Now, $N = (N - S) + S$, so since $3 \mid S$ and $3 \mid (N - S)$, we conclude that $3 \mid N$.

Use this proof as a model for those in the succeeding problems. First write out a proof for 5-digit integers (in Problem 7, use an 8-digit integer), and then write a general proof if you can.

In all "divisibility by M" tests in this chapter, the strategy is the same: For every integer N, find a quantity S such that the quantity $N - S$ always has a factor of M. Thus if $M \mid S$, then you can conclude that $M \mid N$. Why? In most cases I have given you the quantity S so your job is to show that $M \mid (N - S)$.

Problem 12 Examine the relationship between factors of 9, 9876543219, 98765432198765432l9, ..., and 987, 987654321987, 987654321987654321987, ..., etc. for each of the first five odd integers in the set.

Problem 14 Look again at the proof for Problem 2 and see what role is played by the base 10. The key point there was that $3 \mid (10^k - 1)$ for every integer k. For what bases b, then, is it true that $3 \mid (b^k - 1)$ for every integer k? Using algebra, you can show that $b^k - 1$ has $b - 1$ as a factor for any integer k. You have seen this fact illustrated in the factorization

$$b^2 - 1 = (b - 1)(b + 1)$$
$$b^3 - 1 = (b - 1)(b^2 + b + 1)$$
$$b^4 - 1 = (b - 1)(b + 1)(b^2 + 1),$$

so in general $(b - 1) \mid (b^k - 1)$ for any integer k. Thus the question now becomes: For what bases b is $3 \mid (b - 1)$? In this case, 3 will automatically be a factor of $b^k - 1$ for any integer k by the reasoning above.

REFLECTIONS ON PROBLEM SOLVING

II

The first time we paused to reflect, I indicated a four-step procedure for solving problems. These steps were:

(1) Try some numerical examples.
(2) Make a conjecture based on the examples
(3) Check the conjecture with a few more examples.
(4) Verify the conjecture.

Step (4) as I said is the most difficult, so in this section I'm going to try and shed a little light on it by making several suggestions which I've found helpful.

1) After you've tried several numerical examples and made a conjecture, look closely at your examples to see if you can abstract any feature which is common to all of them. This common feature may be the key to the general reasoning required to verify your conjecture. (For example, in Chapter 2, Problem 8, you can abstract the idea that every prime greater than 11 can be written as the sum of 9 plus an even composite integer.)

This is, in fact, the real purpose of looking at examples—to see if you *can* find a general procedure for constructing them (without relying on trial and error), and thus discover the general principle behind the problem.

2) Another closely related suggestion, which has been made before, is to write an algebraic equation or expression representing the problem; and then solve the problem by utilizing your algebraic skills to express the equation (or expression) in such a way that the solution

is apparent. The very fact that you can write down an algebraic equation (or expression) means that you have abstracted a general procedure for constructing examples.

3) Suggestions 1) and 2) also apply to solving problems which require that you prove a given statement true or false, or prove whether the answer to a question is yes or no. Remember, all that is necessary to justify the falsity of a statement or a "no" response is one counterexample. (I will frequently ask for more than one in order to be sure you understand the point to a problem, but from a logical point of view one counterexample is sufficient.)

4) In every problem something is known, something is unknown, and there is a condition relating the known facts to the unknown conclusion. If you find you are having difficulty with a problem, reread it, look at your examples again, and then ask yourself—"Did I make use of *all* the facts I know about the problem?" and "Did I use the entire condition?"

5) The more problems you have solved, the better the use you can make of this suggestion. Try to relate your present problem to one you have solved before: "Does this problem have the same or a similar unknown as one I have previously solved?" "Are these examples I have computed reminiscent of those found in another problem?" "Does this problem have the same or a similar condition as one I have previously solved?" If the answer to any of these questions is "yes", carefully review your work on those earlier problems.

6) If you make a conjecture which turns out to be incorrect, do not discard your work completely. Most guesses or conjectures contain at least a portion of the truth—a step in the right direction. You probably have

solved part of the problem, so now go back and reexamine your examples, and compute some different examples. Many times, the reason you made a false conjecture was that you failed to compute enough examples—you found a pattern which happened to be true for only the examples you had computed, but unfortunately this pattern was not true for all integers. Now look at your new examples, make a new conjecture, check your conjecture with a few more examples, and then verify your conjecture.

7) This last suggestion is simply a psychological truism. Henri Poincaré has described what he feels are the four phases of the thinking process:

(1) The preparatory phase during which you bring together all the background knowledge you feel is necessary to solve the problem, and examine the problem in various ways in an effort to find a solution.

(2) The incubation phase during which you appear to lay the problem aside but at the same time the traces of your preparation seem to be resolving themselves into some sort of meaningful relation.

(3) The illumination phase when you have the flash of insight, the "ah-hah" experience, as the solution to your problem becomes evident.

(4) The verification phase where you sit down once more and systematically test the veracity of your insight.

My suggestion involves (2), the incubation phase. After you feel you have understood the problem and have worked on it a reasonable length of time without success, put the problem aside for awhile. Give your thoughts time to incubate. Then *return to the problem* several hours or several days later. At that time you have a much better chance of success. In fact, it often happens that a solution will occur to you "out of the blue"—while you are involved in some other task, or when you first wake up in the morning.

I hope that by now you are enjoying solving problems. The satisfaction of conceiving an idea, of seeing a sudden light after a period of hesitation and frustration, is a feeling every student of mathematics should have a chance to experience.

All Gaul is divided into three parts
— *J. Caesar*

6 The Division Algorithm

We pause in our discussion of finding prime factorizations long enough to discuss the Division Algorithm, a tool we will use in several of the succeeding chapters.

Previously we have discussed when the integer a is a factor of the integer b, and the fact that every integer has a unique prime factorization. But suppose a is not a factor of b $(a \nmid b)$. Then b lies between two consecutive multiples of a, say qa and $(q+1)a$. For example, $3 \nmid 22$, and 22 lies between $21 = \underline{7} \cdot 3$ and $24 = \underline{8} \cdot 3$. Or, put another way, when b is divided by a, you are asked to subtract the largest multiple of a (qa) which is less than b. Thus for all integers a and b where $a \nmid b$, integer b is a "little more than" some multiple of integer a, say qa, with a difference $(b - qa)$ equal to an integer r. More formally this fact is known as the

Division Algorithm

For any two integers a and b, $a \leq b$, there exist integers q and r such that $b = aq + r$ with $0 \leq r \leq (a-1)$. Clearly, if $a \mid b$, then $r = 0$.

For example, if $b = 172$ and $a = 14$, then $q = 12$ and $r = 4$, so that $172 = (14 \cdot 12) + 4$.

We can graphically represent the Division Algorithm on the number line. If we locate points for a and b, then $b = aq + r$ means there exists an integer q such that if we take q copies of the segment between 0 and a, we get a segment whose length is less than that of the segment between 0 and b. And if we take $q+1$ copies of the segment between 0 and a, we get a segment whose length is greater than that of the segment between 0 and b. Graphically, r is the length of the segment between qa and b.

The Division Algorithm is simply a restatement of the familiar elementary school chant: "The divisor times the quotient plus the remainder equals the dividend."

Problem 1 (a) If $a = 43$ and $b = 574$, find q and r.
(b) If $a = 23$ and $b = 5957$, find q and r.

Problem 2 Explain carefully why $0 \leq r \leq a - 1$.

Problem 3 Find all integers which, when
(a) 100 (b) 253 (c) 500 (d) 1518
are divided by them, yield a remainder $r = 1$. Explain your reasoning.

Problem 4 For each of the following, find five examples of integers b such that
(a) $r = 3$ when $a = 4$ *and* (simultaneously) $r = 1$ when $a = 6$,
(a) $r = 4$ when $a = 5$ *and* (simultaneously) $r = 1$ when $a = 10$,
(c) $r = 2$ when $a = 5$ *and* (simultaneously) $r = 3$ when $a = 7$,
(d) $r = 4$ when $a = 6$ *and* (simultaneously) $r = 8$ when $a = 9$,
or explain why none exist.

Problem 5

(a) What values are possible for r when b is a prime greater than 6 and $a = 6$.

(b) What values are possible for r when b is a prime greater than 30 and $a = 30$.

Explain your answer in each case.

Problem 6

(a) Find the unit's digit of 3^{999}; of $175{,}843^{999}$.

(b) Find the last two digits of 7^{999}; of 3^{999}.

Problem 7

(a) Find the unit's digit of 2^{10}; of $175{,}482^{10}$.

(b) Find the last two digits of $175{,}482^{10}$.

Problem 8 [First proposed by Pierre Fermat, a famous French mathematician, in 1635] Find r if $b = 5^{999{,}999}$ and $a = 7$.

Another chance to write a proof is contained in

Problem 9 Suppose a and b are both multiples of some integer k. Show that r is also a multiple of k. Illustrate this idea with an example when $k = 5, 9, 17$.

Problem 10

(a) For what (type of) integers n do n and n^2 have the same remainder when divided by 4?

(b) For what (type of) integers n do n and n^3 have the same remainder when divided by 6? by 12?

Explain your reasoning in each case.

Next is a problem [part (c) of Problem 11] which reportedly appeared in a volume of Ripley's *Believe It or Not*—with an incorrect answer!

Problem 11

(a) An integer when divided by 2 leaves a remainder of 1, and when divided by 3 leaves a remainder of 2. Find the smallest such integer.

(b) An integer when divided by 2 leaves a remainder of 1, when divided by 3 leaves a remainder of 2, when divided by 4 leaves a remainder of 3, and when divided by 5 leaves a remainder of 4. Find the smallest such integer.

(c) An integer when divided by 10 leaves a remainder of 9, when divided by 9 leaves a remainder of 8,..., when divided by 3 leaves a remainder of 2, and when divided by 2 leaves a remainder of 1. Find the smallest such integer.

Problem 12 Let a, b, c, d denote any four integers. Prove that $12 \mid (a-b)(b-c)(c-d)(d-a)(a-c)(b-d)$. (This product may be negative or zero.)

In Chapter 12 we will look at more sophisticated problems involving the Division Algorithm.

Now it's your turn to formulate a problem, specifically:

Problem 13 Formulate, and solve if you can, a problem involving the Division Algorithm.

HINTS AND COMMENTS

Problems 6–8 Another common tool of problem solvers is utilized in these three problems: To find r for $b = 5^{999,999}$ seems difficult because $5^{999,999}$ is such a large number; so first try to answer a simpler problem; i.e. find r if $b = 5$, if $b = 5^2$, if $b = 5^3$, . . . , and keep doing this until you find a pattern.

In Problem 6(b)—persevere!

Problem 9 Refer to Chap. 1, Prob. 12 (page 12), for a model. We know three things: $b = aq + r$, $b = xk$, $a = yk$ (for some integers x and y). We wish to find an integer z such that $r = zk$. Find z.

Problem 10 If n and n^2 have the same remainder when divided by 4, then

$$n^2 = 4q_1 + r$$
$$n = 4q_2 + r,$$

so that $n^2 - n = 4(q_1 - q_2)$. So now decide for what (type of) integers n is $4 \mid (n^2 - n)$.

Problem 11(b) Let N be the desired integer:

$$\begin{aligned} N &= 2q_1 + 1 \\ &= 3q_2 + 2 \\ &= 4q_3 + 3 \\ &= 5q_4 + 4. \end{aligned}$$

Look at the integer $N + 1$.

Problem 12 Consider what happens when a, b, c, d are divided by 3, and by 4.

REFLECTIONS ON PROBLEM SOLVING

III

In this final discussion of problem solving, I would like to indicate what I feel are some of the major faults of beginning problem solvers. Hopefully, if you recognize any of these faults in yourself, you can then improve your problem-solving ability by remedying them.

(1) As with so many other tasks, an essential ingredient in the success of solving a problem is the desire to solve it. The old football cliché that desire is 50% of performance is true in problem solving as well. Consequently, I feel the major fault (if you can call it a fault) of problem solvers, is the lack of real determination to solve the problem.

(2) Failure to completely understand the problem is another common error of problem solvers. This lack of understanding could be psychological—failure to completely concentrate on the problem. Or it could be technical—failure to compute enough examples to see what the problem is all about.

(3) Another frequent fault is the failure to *really examine* your numerical examples carefully—analyze precisely what you did when you carried out the computations. You probably should compute more examples than you are presently doing. You should try a few large integers—many times, the inability to carry out a computation for a large integer will point out the crux of the problem.

(4) The lack of patience, care, perseverance, or whatever you wish to call it, is another shortcoming of many beginning problem solvers. While I freely admit that trying to solve a certain problem can be a frustrating experience, you can reduce this frustration by taking the time to carefully check each step in your work. I have often seen students needlessly frustrated because they computed an example incorrectly, copied something incorrectly, misread the problem, made an algebraic error, made a basic error in reasoning, etc.

Many students completely give up on a problem if they cannot solve it in 30 minutes. Don't give up! Put your problem aside and return to it later. As I've said before—give your thoughts a chance to incubate.

(5) The failure to check your result to see whether your result is reasonable is a frequent error. You may be so glad to get *any* result that you throw down your pencil and forget it—no result, however implausible, would shock you. For example in Chapter 1, Problem 13, if you *could* construct a proof that if $a \mid c$ and $b \mid c$, then $ab \mid c$, you should be shocked because this says that you can take any two factors of an integer, multiply them together, and the product would always be factor of the integer. In particular, take $a = b = c$ in the problem. There your proof would imply that $a^2 \mid a$!

This ends my discussion of problem solving. Probably not everything I've said has meaning for you. This is understandable because no two problem solvers work in precisely the same way. But by this time I hope you now enjoy problem solving and have begun to develop your own style. Above all, I hope you agree that problem solving is important. As George Polya (my favorite mathematical author) so convincingly argues:

Solving problems is the specific achievement of intelligence, and intelligence is the specific gift of man. The ability to go round an obstacle, to undertake an indirect course when no direct course presents itself, raises the clever animal above the dull one, raises man far above the most clever animals, and men of talent above their fellow men.

Nothing is more interesting for us humans than human activity. The most characteristically human activity is to solve problems, thinking for a purpose, devising means to some end. Our aim is to understand this activity—it seems to me that this aim deserves a good deal of interest.

7

The most common things are the most useful.

— *W. Penn*

Greatest Common Factor

If I were forced to give a one-sentence description of the content of the first six chapters, I would say: "Given any integer n, say what you can about those integers which are factors of n, and perhaps a few words about those which are not." In this and the next chapter we are going to look at similar problems with regard to pairs of integers and, on occasion, triples or quadruples of integers. We can compare of our activity to that of chemists (as we did in Chapter 3); given two (or three or four) elements, how can we combine them to form compounds?

Look at a pair of integers, (a, b). Since we are concerned with factors, (at least) two questions arise:

(1) What is the largest integer which is a factor of both a and b?

(2) What is the smallest integer of which both a and b are factors?

For example, if $a = 48$ and $b = 72$, the answer to question (1) is 24, while the answer to (2) is 144. Chapter 7 is devoted to examining (1), while in Chapter 8, question (2) will be discussed.

Formally, we say that the *greatest common factor* of two integers a and b [notation $\text{GCF}(a, b)$] is the largest integer which is a factor of both a and b. One way of finding $\text{GCF}(a, b)$ is to find the prime factorizations of a and b, and from them determine the greatest common factor: e.g. $3900 = 2^2 \cdot 3 \cdot 5^2 \cdot 13$ and $4290 = 2 \cdot 3 \cdot 5 \cdot 11 \cdot 13$, so $\text{GCF}(3900, 4290) = 2 \cdot 3 \cdot 5 \cdot 13 = 390$. (Exactly how was this integer computed?) In like manner, $\text{GCF}(3300, 4680) = 60$ since $3300 = 2^2 \cdot 3 \cdot 5^2 \cdot 11$ and $4680 = 2^3 \cdot 3^2 \cdot 5 \cdot 13$, so $\text{GCF}(3300, 4680) = 2^2 \cdot 3 \cdot 5 = 60$. If $\text{GCF}(a, b) = 1$, the integers a and b are said to be *relatively prime* (i.e. they share no common prime factors). For example, (6, 13) and (10, 21) are two pairs of relatively prime integers.

Problem 1 Find GCF(a, b) if
(a) $a = 54$, $b = 72$
(b) $a = 198$, $b = 162$
(c) $a = 2^3 \cdot 3^2 \cdot 5^4 \cdot 7^3 \cdot 11^2$, $b = 2^2 \cdot 3^3 \cdot 5^3 \cdot 7 \cdot 11^3$
(d) $a = 23$, $b = 0$ (Consider 0 to be an integer here.)

Problem 2 Under what condition does GCF(a, b) = a?

Problem 3 If GCF(a, b) = 3, what values are possible for
(i) GCF(a^2, b)
(ii) GCF(a^3, b)
(iii) GCF(a^2, b^3)?

Problem 4 Find all pairs of integers (x, y) satisfying each of the following:
(a) $xy = 2325$ and GCF(x, y) = 5,
(b) $xy = 90{,}000$ and GCF(x, y) = 5,
(c) $xy = 90{,}000$ and GCF(x, y) = 10,
(d) $xy = 225{,}220$ and GCF(x, y) = 5.

Problem 5 The GCF(a, b) can be viewed as the largest element contained in the intersection of two sets (a method used in many elementary school texts). Let $A = \{$*all* factors of $a\}$, $B = \{$*all* factors of $b\}$, (arranging the elements of A and B in ascending order). Then $A \cap B = \{$all common factors of a and $b\}$ so that GCF(a, b) is the largest element in $A \cap B$. For example, if $a = 15$ and $b = 20$, then $A = \{1, 3, 5, 15\}$ and $B = \{1, 2, 4, 5, 10, 20\}$, so $A \cap B = \{1, 5\}$, hence GCF(a, b) = 5. If $a = 54$ and $b = 72$, find A, B, $A \cap B$, and GCF(a, b). Do the same for $a = 60$, $b = 84$.

Problem 6 The greatest common factor of more than two integers (say $a, b, c, d, \ldots$) is defined (as you might suspect) as the largest integer which is a factor of each of the integers $a, b, c, d, \ldots,$ and is denoted by GCF$(a, b, c, d, \ldots)$.
Find: (a) GCF(54, 72, 108)
(b) GCF(54, 72, 99)
(c) GCF$(2^3 \cdot 3^2 \cdot 5^4 \cdot 7^5,\ 2 \cdot 3^4 \cdot 5^5 \cdot 7^4,\ 2^2 \cdot 3^3 \cdot 5^3 \cdot 7^5)$
(d) GCF(54, 72, 0) [Consider 0 to be an integer here.]

Problem 7

(a) If GCF(a, b) = GCF(a, c) = GCF$(b, c) = 1$, does GCF$(a, b, c) = 1$? Justify your answer.

(b) Conversely, if GCF$(a, b, c) = 1$, does GCF(a, b) = GCF(b, c) = GCF$(a, c) = 1$? Justify your answer.

(c) If GCF$(a, b) = 1$ and GCF$(b, c) = 1$, does GCF$(a, c) = 1$? Justify your answer.

Problem 8

(a) If $a = b + c$, what is the relationship between GCF(a, b) and GCF(b, c)? Justify.

(b) If $b = aq + r$, what is the relationship between GCF(a, b) and GCF(a, r)? Justify.

While finding the greatest common factor of two integers by finding their prime factorizations is easy if the integers are fairly small, for large integers this process would be lengthy, to say the least. For this reason we seek an alternative process—one which is suggested by Problem 8(b).

If $b = aq + r$, then you have probably conjectured that GCF(b, a) = GCF(a, r), but (the inevitable question) why? To save writing, let d_1 denote GCF(b, a) and d_2 denote GCF(a, r).

Now, $d_1 \mid b$ and $d_1 \mid aq$, and thus $d_1 \mid (b - aq)$, i.e. $d_1 \mid r$. (See Problem 9, Chapter 6.) That is, d_1 is a common factor of a and r, so $d_1 \leq d_2$; i.e. d_1, a common factor of a and r, is less than or equal to d_2, the *greatest* common factor of a and r. Similarly, $d_2 \mid a$ and $d_2 \mid aq$, and also $d_2 \mid r$, so that $d_2 \mid aq + r)$, i.e. $d_2 \mid b$. (See Problem 12, Chapter 1, page 00.) Thus, reasoning as before, $d_2 \leq d_1$. Together, we have $d_1 \leq d_2$ and $d_2 \leq d_1$, so $d_1 = d_2$.

Think about what this says! To find GCF(b, a) we can instead find GCF(a, r), which is easier since a and r are smaller integers.

Example 1. Find GCF(14,673, 4453). Now, we find that

(1) 14,673 = (4453·3) + 1314,

so GCF(14,673, 4453) = GCF(4453, 1314). But these integers are still rather large, so we repeat the process: we divide 4453 by 1314 obtaining:

(2) 4453 = (1314·3) + 511,

so GCF(4453, 1314) = GCF(1314, 511). Still too large. Do it again!

(3) 1314 = (511·2) + 292

and again

(4) 511 = (292·1) + 211

and again

(5) 292 = (211·1) + 73

and one final time

(6) 219 = (73·3) + 0.

Hence applying the result of Problem 8(b) six times we have:

GCF(14,673, 4453) = GCF(4453, 1314) = GCF(1314, 511)
= GCF(511, 292) = GCF(292, 219)
= GCF(219, 73) = GCF(73, 0).

But GCF(73, 0) = 73, so GCF(14,673, 4453) = 73.

A simple way to keep track of these computations is the following:

```
                 3
(1)   4453 |14,673
            13,359      3
      (2)   1,314 |4453
                   3942      2
            (3)  511 |1314
                      1022      1
               (4)    292 |511
                           292      1
                  (5)  219 |292
                             219      3
                     (6)    73 |219
                                219
                                  0
```

GCF (14,673, 4453) = 73, reasoning as above. This process of finding the GCF(a, b) is commonly referred to as the *Euclidean Algorithm.*

Example 2. We apply the Euclidean Algorithm to our previous example—to find GCF(4290, 3900):

(1) 4290 = (3900·1) + 390

(2) 3900 = (390·10) + 0.

So GCF(4290, 3900) = GCF(3900, 390) = GCF(390, 0) = 390.

```
          1
3900 |4290
      3900    10
      390 |3900
           3900
              0
```

Problem 9 Using the Euclidean Algorithm, find GCF(b, a) if:

(a) $b = 84$, $a = 18$
(b) $b = 1{,}326$ $a = 650$
(c) $b = 7{,}469$, $a = 2{,}464$
(d) $b = 4{,}001$, $a = 2{,}689$
(e) $b = \underbrace{1{,}111, \ldots, 111}_{\text{100 times}}$, $a = 11{,}111{,}111$.

Problem 10 Find all possible values for:

(a) GCF(p, q)	p, q distinct primes
(b) GCF$(n, n + 1)$	n any integer
(c) GCF$(n, n + 2)$	n any integer
(d) GCF$(n, n + 5)$	n any integer
(e) GCF$(n, n + 30)$	n any odd integer
(f) GCF$(n, n + 30)$	n any integer
(f) GCF$(n^2 + 1, n + 1)$	n any integer
(g) GCF$(8n + 7, 5n + 6)$	n any integer

Justify your answers.

Problem 11 For what (type of) integers n is

(a) GCF$(n + 6, n + 2) = 4$?
(b) GCF$(6, n + 3) = 3$?

Justify your answer in each case.

Problem 12 Consider the sequence of integers
1, 1, 2, 3, 5, 8, 13, 21, 34.

(a) Write down the next six terms of this sequence.

(b) What is the greatest common factor of any two consecutive terms? Why?

Problem 13 Let a and b be any two integers, with $a > b$, for which GCF(a, b) = 1. What values are possible for GCF($a + b$, $a - b$)? Justify your answer.

Problem 14 Consider the three integers 2718, 3875, and 4486. Find the largest integer n which yields the same remainder in each case when each of the three integers is divided by n.

To close this chapter, here is a problem which you have almost done before in another context.

Problem 15 What is the largest integer n for which the following statement is true: "All integers ($\neq 1$) less than n and relatively prime to n are primes."

For example, the statement is true for $n = 12$ since 5, 7, 11 are all integers less than 12 and relatively prime to 12, and they are primes; however, the statement is false for $n = 10$ since 9 is an integer less than 10 and relatively prime to 10, but 9 is *not* prime.

Recall that two integers a and b are *relatively prime* if GCF(a, b) = 1.

HINTS AND COMMENTS

Problems 10 and 13 Use the first step of the Euclidean Algorithm; e.g. in part (c), $(n + 2) = n + 2$, so $\mathrm{GCF}(n + 2, n) = \mathrm{GCF}(n, 2)$. Hence the only possible values to consider are 1 and 2. This means if we can find one value of n for which $\mathrm{GCF}(n, 2) = 1$, then 1 is a possible value. Similarly for 2. Clearly, for any odd integer n, we have $\mathrm{GCF}(n, 2) = 1$, and for any even integer n, $\mathrm{GCF}(n, 2) = 2$. Thus 1 and 2 are all possible values.

In Problem 10(g) use more than one step of the Euclidean Algorithm.

Problem 12 This sequence of integers is known as the *Fibonacci sequence*, named after the 13th century Italian mathematician Leonardo of Pisa (Fibonacci was his stage name). The sequence arises in connection with the following problem (which he formulated in 1202):

"What is the number of pairs of rabbits at the beginning of each month if a single pair of newly born rabbits is put into an enclosure at the beginning of January, and if each pair breeds a new pair at the beginning of the second month following birth and an additional pair at the beginning of each month thereafter?"

For part (b), look again at Problem 8(a) in this chapter.

Problem 14

$$2718 = nq_1 + r$$
$$3875 = nq_2 + r$$
$$4486 = nq_3 + r.$$

Look at $(3875 - 2718)$ and $(4486 - 3875)$.

Problem 15 Review Chapter 6.

The most common things are still the most useful.

— *W. Penn Jr.*

8 Least Common Multiple

A companion concept to the greatest common factor, which answers question (2) posed in the previous chapter, is the *least common multiple* of two integers a and b, which is defined as the smallest integer containing both a and b as factors. To the surprise of no one, it is denoted by LCM(a, b). For example, LCM$(9, 6) = 18$. The LCM(a, b) can also be determined easily by looking at the prime factorizations of a and b. To find LCM(4290, 3900), we write

$$4290 = 2 \cdot 3 \cdot 5 \cdot 11 \cdot 13$$
$$3900 = 2^2 \cdot 3 \cdot 5^2 \cdot 13,$$

so LCM$(4290, 3900) = 2^2 \cdot 3 \cdot 5^2 \cdot 11 \cdot 13 = 42{,}900$. (Exactly how was this integer computed?)

Problem 1 Find LCM(a, b) if:

(a) $a = 30, \quad b = 28$

(b) $a = 54, \quad b = 72$

(c) $a = 29{,}347, \quad b = 2$

(d) $a = 2^3 \cdot 3^2 \cdot 5^4 \cdot 7^5 \cdot 13, \quad b = 2^2 \cdot 3^3 \cdot 5^3 \cdot 17.$

Problem 2 Under what condition is LCM$(a, b) = $ b?

Problems 3 Show how you can interpret LCM(a, b) as the smallest element in the intersection of two sets. Illustrate with an example. (See Problem 5, Chapter 7, page 61, before you begin.)

Problem 4 The least common multiple of more than two integers, say $a, b, c, d, \ldots$, is defined as the smallest integer of which each of the integers $a, b, c, d, \ldots$ is a factor, and is denoted by $\text{LCM}(a, b, c, d, \ldots)$. Find:

(a) LCM(30, 28, 105)
(b) LCM(54, 72, 99)
(c) LCM(4; 5; 9; 11; 6,776,538,120)

Problem 5 Determine whether the following statements are true or false. Justify your answers. For all integers a, b, c:

(a) If $\text{LCM}(a, b) = \text{LCM}(a, c)$, then $b = c$.
(b) If $\text{GCF}(a, b) = \text{GCF}(a, c)$, then $\text{LCM}(a, b) = \text{LCM}(a, c)$.
(c) $\text{LCM}(a^2, b^2) = [\text{LCM}(a, b)]^2$.

Problem 6 Find all pairs of integers (a, b) for which

(a) $\text{GCF}(a, b) = 2$ and $\text{LCM}(a, b) = 24$
(b) $\text{GCF}(a, b) = 6$ and $\text{LCM}(a, b) = 72$
(c) $\text{GCF}(a, b) = 36$ and $\text{LCM}(a, b) = 2160$.

Look again at your answer to Problem 6. You will then discover that

Problem 7 For any two integers (a, b), $\text{GCF}(a, b) \mid \text{LCM}(a, b)$. Prove this.

To "tie together" the two concepts of GCF and LCM, I now ask you to prove

Problem 8 For any two integers (a, b), $\text{LCM}(a, b) \cdot \text{GCF}(a, b) = a \cdot b$.

If a and b are large integers, Problem 8 gives an efficient way of computing LCM(a, b) by setting LCM$(a, b) = a \cdot b/$GCF(a, b) and computing GCF(a, b) using the Euclidean Algorithm.

Problem 9 Find LCM(899, 1073).

Some other applications of Problem 8 are contained in the next three problems.

Problem 10

(a) How many multiples of 15 are there in the sequence 10, 20, 30,..., 140, 150?

(b) How many multiples of 24 are there in the sequence 15, 30, 45, 60,..., 345, 360?

(c) In general, how many multiples of n are there in the sequence m, $2m$, $3m$,..., $(n - 1)m$, nm? Explain your reasoning.

Problem 11 Find a general formula for

(a) LCM(p, q) if p, q are primes

(b) LCM$(n, n + 1)$ if n is any integer

(c) LCM$(n, n + 2)$ if n is any odd integer

(d) LCM$(n, n + 2)$ if n is an even integer

(e) LCM$(n, n + 6)$ if n is an odd multiple of 3

Justify your answers.

Problem 12 For what (type of) integers n is LCM$(6, n + 3) = 2n + 6$? Justify your answer.

In this problem, you will investigate whether Problem 8 remains true for three integers a, b, c. Specifically:

Problem 13

(a) Find all triples of integers (a, b, c) for which $GCF(a, b, c) = 10$ and $LCM(a, b, c) = 1{,}000$.

(b) Under what condition does $GCF(a, b, c) \cdot LCM(a, b, c) = abc$?
Prove your answer.

Problem 14 Formulate and solve a problem relating the GCF of a set of two or more integers to the LCM of the set.

HINTS AND COMMENTS

I don't really think any hints are necessary for this chapter. The only comment is to be sure you understand *why* $GCF(a, b) \cdot LCM(a, b) = ab$ for any two integers. Most of the problems in this chapter are applications of this idea.

9

Count your blessings.

Counting the Factors

Throughout these chapters we have looked at the factors of an integer in several contexts, most notably in the Fundamental Theorem of Arithmetic and in the notion of greatest common factor of two or more integers. But thus far we have evaded the question "How many factors does an arbitrary integer have?" For example, in Chapter 2 you saw that primes were integers having exactly two factors.

Let $d(n)$ denote the number of factors of the integer n (including 1 and n). Thus $d(22) = 4$ since 1, 2, 11, 22 are all factors of 22; and $d(p) = 2$ for any prime p.

Students have always enjoyed this next set of problems—I hope you do too.

Problem 1 Complete the following table:

n	2	3	4	5	6	7	8	9	10	11	12	15	16	18	24	25	27	30	36
$d(n)$																			

Problem 2 Find $d(p^2)$, $d(p^3)$, $d(p^4)$, $d(p^e)$ where p is any prime, e any integer,

Problem 3 Find $d(pq)$, $d(p^2q)$, $d(pq^2)$, $d(p^2q^2)$, $d(p^3q)$ where p, q are any two distinct primes. In each case try a few examples first.

Problem 4 Given the prime factorization $n = p_1^{e_1} p_2^{e_2} \cdots p_k^{e_k}$, find $d(n)$ in terms of e_i, $i = 1, 2, \ldots, k$. (That is, knowing the exponents of all primes in the prime factorization of an integer n, how could you compute $d(n)$?)

Problem 5 Find (a) $d(42)$ (b) $d(240)$ (c) $d(2^{19}3^{9}5^{29})$, using the formula you found in Problem 4.

Problem 6 Complete the following table. In each case find the smallest possible integer n having $d(n)$ factors.

$d(n)$	2	3	4	5	6	7	10	12	24	28	36
n											

Problem 7 For any arbitrary integer k, does there always exist an integer n such that $d(n) = k$? Justify your answer.

Problem 8 For what (type of) integers n is $d(n)$ a prime? Justify.

Problem 9 For what (type of) integers n is $d(n)$ an odd integer? Justify.

Problem 10 How must the integers n and m be related so that $d(n)d(m) = d(nm)$? First find several examples where $d(n)d(m) \neq d(nm)$. If $d(n) \neq d(nm)$, which side of the inequality is larger? Why?

Problem 11 Find the number of positive integers that leave a remainder of 4 when divided *into*
(a) 100 (b) 200 (c) 600 (d) 1001.

Problem 12 Let $e(n)$ denote the number of even factors of n. For example, $e(12) = 4$ since 2, 4, 6, 12 are the even factors of 12.
(a) Find $e(6)$, $e(16)$, $e(30)$, $e(60)$.
(b) 1 is a factor of all integers, but what integers have all even factors except 1; i.e. for what (type of) integers n is $e(n) = d(n) - 1$? Explain your reasoning.
(c) Give several examples of integers, exactly half of whose factors are even. In general, for what (type of) integers n is $2e(n) = d(n)$.
(d) Formulate and solve another problem involving a relationship between $e(n)$ and $d(n)$.

Problem 13 In how many ways can
(a) 36 (b) 60 (c) 288 (d) 210
be written as the product of two relatively prime integers? In general, for any integer n, how many ways can n, be written as the product of two relatively prime integers? Justify your answer if you can.

Problem 14 For what (type of) integers n is $d(n)$ less than $2\sqrt{n}$? Justify your answer.

This last problem is conceptually the same as one of the previous fourteen.

Problem 15 Imagine 1000 lockers, all closed, and 1000 men. Suppose the first man goes along and opens every locker. Then the second man goes along and shuts every other locker. Suppose the third man goes along and changes the state of every third locker (if it's open, he shuts it, and vice versa). Suppose the fourth man goes along and changes the state of every fourth locker, etc. until all 1000 men have passed by all the lockers. Which lockers are open at the end of this game?

HINTS AND COMMENTS

Problem 4 In Problem 2 you hopefully discovered that $d(p^e) = e + 1$. So then

$$d(p_1^{e_1} p_2^{e_2} \cdots p_k^{e_k}) = (e_1 + 1)(e_2 + 1) \cdots (e_k + 1);$$

any factor of n is the product of factors of each of the $p_i^{e_i}$. Explain this statement in your own words to be sure you understand it before proceeding to the next problems.

Problem 6 Be careful on $d(n) = 12, 24, 28, 36$.

Problem 11 Use the Division Algorithm.

Problem 14 If $a \mid n$ and $1 \le a \le \sqrt{n}$, then $(n/a) \mid n$, $n/a \ge \sqrt{n}$. This problem utilizes the same principle as Problem 3, Chapter 2, page 18.

No perfection is so absolute,
that some impurity doth not pollute.

W. Shakespeare

10

Perfect Numbers

To the ancients, science and mathematics were often intermingled with philosophy and metaphysics; mathematical objects (e.g. numbers) had personalities all their own. It comes then, as no surprise that the ancients called a number *perfect* if it was an integer equal to the sum of its proper factors, or, put another way, a number is perfect if the sum of all factors of the number equals twice the number. For example, $6 = 1 + 2 + 3$ is the smallest perfect number, since 1, 2, 3 are the only proper factors of 6, or $12 = 1 + 2 + 3 + 6$. The fact that in 1644 only eight perfect numbers were known prompted Mersenne, a French mathematician, to write: "We see clearly from this fact how rare are perfect numbers and how right we are to compare them to perfect men." As late as 1971 only 16 additional perfect numbers, all very large, had been found. The largest known perfect number is $2^{19,937}(2^{19,937} - 1)$, an integer having 12,003 digits. Even though very few perfect numbers are known, analyzing their properties is still a valuable exercise for students striving to perfect their mathematical talents.

The study of perfect numbers, then, is really the study of the sum of the factors of an integer. So let's study just that. For any integer n let $\sigma(n)$ denote the sum of all factors of n. (σ, read "sigma," is the Greek letter s.) Thus $\sigma(22) = 1 + 2 + 11 + 22 = 36$.

Problem 1 Complete the following table:

n	2	3	4	6	8	9	12	18	20	27	28	30	36
$\sigma(n)$													

Problem 2 Do there exist pairs of integers (m, n) for which $\sigma(m) = \sigma(n)$? Either find three such pairs or show that none exist.

Problem 3 Verify that there is no integer n for which $\sigma(n) = 2$, $\sigma(n) = 5$. For every integer $k > 6$, does there exist an integer n for which $\sigma(n) = k$? Show that the conclusion is always true or find three examples where it is false.

Problem 4 In Problem 1, there were a few integers n for which $\sigma(n)$ was odd. Describe a set of infinitely many integers n for which $\sigma(n)$ is odd. Prove your answer.

Problem 5 For what (type of) integers n is $\sigma(n)$ greater than $n + \sqrt{n}$? Justify.

We can define a perfect number in terms of $\sigma(n)$ as any integer n for which $\sigma(n) = 2n$.

Problem 6 Find all perfect numbers less than 30, and show that $496 = 2^4(2^5 - 1)$ is a perfect number.

Problem 7 Suppose p is a prime. Show that if $2^p - 1$ is prime, then $2^{p-1}(2^p - 1)$ is a perfect number.

Problem 8 What unit's digits are possible for even perfect numbers? Justify your answer.

As Problem 7 suggests, all known perfect numbers are even. The existence or lack of existence of an odd perfect number is another of the classical unsolved problems of mathematics.

Problem 9 A perfect number is one for which the sum of the factors of the number equals twice the number. Suppose we replace the word "sum" by "product." Find all integers less than 30 such that the *product* of all proper factors of n equals n, or, put another way, the product of all factors of n equals n^2.

Problem 10 In general, for what (type of) integers n does the product of all factors of n equal n^2? Justify your answer if you can.

Since so few integers are perfect, we conclude that for most integers, the sum of the proper factors is either more or less than the integer. Since non-perfect integers are so plentiful, we assign the more mundane terms *abundant* and *deficient* to the integers, i.e. an integer n is abundant (deficient) if the sum of the proper factors of n is greater than (less than) n, or in symbols, n is abundant if $\sigma(n) > 2n$, and n is deficient if $\sigma(n) < 2n$.

Problem 11 Classify all integers $n \leq 24$, $n \neq 6$, as either abundant or deficient. For example, 4 is deficient since $(1 + 2) < 4$, while 12 is abundant since $(1 + 2 + 3 + 4 + 6) > 12$.

Problem 12 Show that all integers of the form $n = pq$ (p, q distinct primes; $pq \neq 6$) are deficient.

Problem 13

(a) Is the product of two deficient numbers always deficient?

(b) Is the sum of two deficient numbers always deficient?

In each case, show that the conclusion is always true or find three examples where it is false.

Problem 14

(a) Is the product of two abundant numbers always abundant?

(b) Is the sum of two abundant numbers always abundant?

In each case, show the conclusion is always true or find three examples where it is false.

Problem 15 Are all odd integers deficient? Justify.

Problem 16 What type of number (abundant, deficient, or perfect) is

(a) a factor (b) a multiple of a perfect number?

Justify your answer.

HINTS AND COMMENTS

Problem 6 Look at Problem 14, Chapter 9, page 74.

Problem 7 We want to show that if $n = 2^{p-1}(2^p - 1)$ and $2^p - 1$ is prime, then $\sigma(n) = 2^p(2^p - 1)$. What are the factors of n? They are $1, 2, 2^2, 2^3, \ldots, 2^{p-1}$, and $1(2^p - 1), 2(2^p - 1), \ldots, 2^{p-1}(2^p - 1)$ since $2^p - 1$ is prime. Thus

$$\begin{aligned}\sigma(n) &= (1 + 2 + 2^2 + \cdots + 2^{p-1}) + 1 \cdot (2^p - 1) + 2^2(2^p - 1) \\ &\qquad + \cdots + 2^{p-1}(2^p - 1) \\ &= (1 + 2 + \cdots + 2^{p-1}) + (2^p - 1)(1 + 2 + 2^2 + \cdots + 2^{p-1}) \\ &= (1 + 2 + 2^2 + \cdots + 1^{p-1})(1 + 2^p - 1) \\ &= (1 + 2 + \cdots + 2^{p-1})(2^p).\end{aligned}$$

So the proof will be complete if we can show that $1 + 2 + 2^2 + \cdots + 2^{p-1} = 2^p - 1$. You do this part by multiplying $(2 - 1)(1 + 2 + 2^2 + \cdots 2^{p-1})$.

Problem 8 Every even perfect number has the form $2^{p-1}(2^p - 1)$, by Problem 7. First examine the cases $p = 2$ and $p = 5$. Then, in turn, examine what happens when p has the form $10k + 1$, $10k + 3$, $10k + 7$, and $10k + 9$. See Chapter 6, Problems 6 and 7, page 00.

Problem 10 First see if you can find a formula for the product of all factors of n. Keep in mind that if d is a factor of n, then n/d is a factor of n and $d(n/d) = n$.

Problem 15 Conclusions based on looking only at small integers can be wrong.

What percent of integers do you think are abundant? Recent computations have shown that this percent lies between 24.4% and 30.1%. The exact percent is unknown.

11

> I've got your number.
>
> *F. Sinatra*

Geometric Numbers

The ancient Greeks were very interested in geometry (e.g. Euclid), so it comes as no surprise that they tried to study properties of numbers by looking at their geometric representations. In the first part of this chapter you will use properties of some familiar geometrical figures (e.g. squares, triangles) to discover several arithmetical formulas. Your justification for the problems in this section should depend heavily on the geometrical figures—you should see how the arithmetical formula is related to the geometric figure. In the second part of this chapter, you will apply the formulas you have already developed to solve other problems by primarily algebraic means. (One might say that you have solved most of your problems up to now using algebraic means, i.e. you have worked mainly with algebraic expressions and equations.)

(A) Square Numbers

Squares are those numbers which can be represented by a square array of dots. Denote by S_n the nth square number. Thus

$S_1 = 1$ $\qquad$ $S_2 = 4 = 2^2$ $\qquad$ $S_3 = 9 = 3^2$ $\qquad$ $S_4 = 16 = 4^2$

Problem 1 Examine the following partitioning of square numbers.

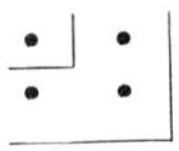

$1 + 3 = 2^2$

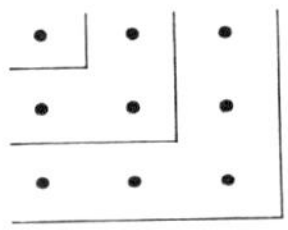

$1 + 3 + 5 = 3^2$

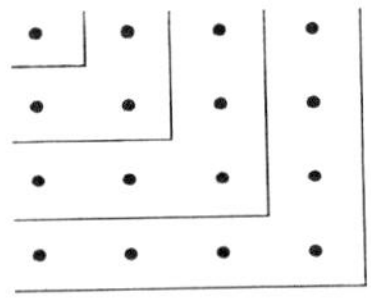

$1 + 3 + 5 + 7 = 4^2$

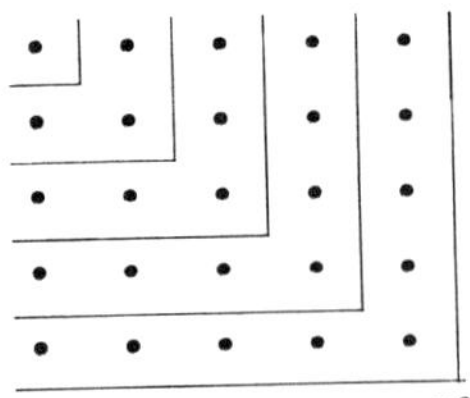

$1 + 3 + 5 + 7 + 9 = 5^2$

(a) Compute $1 + 3 + 5 + 7 + 9 + 11$, $1 + 3 + 5 + \cdots + 21$, $1 + 3 + 5 + \cdots + 97 + 99$.

(b) Now find a general formula for the sum of the first n odd integers. Justify your formula.

(B) Oblong Numbers

Numbers which can be pictured as a rectangular array of dots whose length is one unit more than its width are called *oblong numbers.* Denote by O_n the nth oblong number.

$O_1 = 2$

$O_2 = 6$

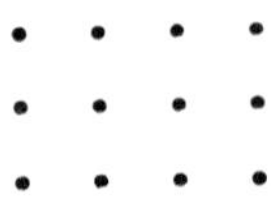

$O_3 = 12$

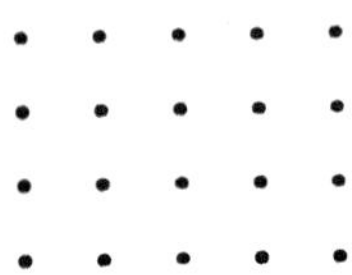

$O_4 = 20$

Problem 2 Compute O_5, O_6, O_8, O_{10}. Then find a general formula for O_n.

Problem 3 Examine the following partitioning of oblong numbers.

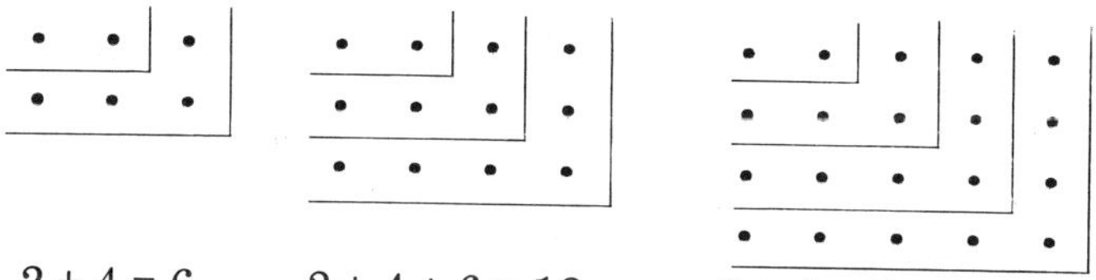

$2 + 4 = 6$ $\quad$ $2 + 4 + 6 = 12$ $\quad$ $2 + 4 + 6 + 8 = 20$

(a) Compute $2 + 4 + 6 + 8 + 10 + 12$, $2 + 4 + 6 + \cdots + 20$, $2 + 4 + 6 + \cdots + 98 + 100$.

(b) Now find a general formula for finding the sum of the first n even integers. Justify your formula.

(C) Triangular Numbers

Triangular numbers are those which can be represented by a (right) triangular array of dots. Denote by T_n the nth triangular number.

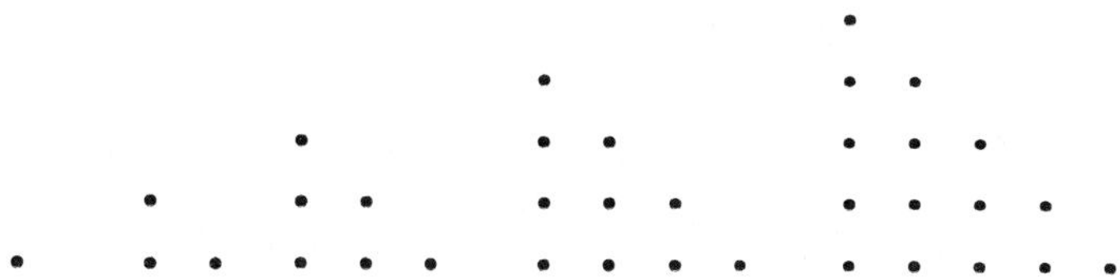

$T_1 = 1$ $\quad$ $T_2 = 3$ $\quad$ $T_3 = 6$ $\quad$ $T_4 = 10$ $\quad$ $T_5 = 15$

Problem 4 Compute T_6, T_8, T_{10}, T_{15}. (Every pool player should know T_{15}.) Then find a general formula for $T_n = 1 + 2 + 3 + \cdots + (n - 1) + n$. Justify your formula.

Problem 5 Use your formula for T_n to compute the following sums:

(a) $1 + 2 + 3 + \cdots + 1000$.
(b) $3 + 6 + 9 + \cdots + 999$.
(c) $5 + 10 + 15 + 20 + \cdots + 1000$.

Problem 6 Find one or more relationships between triangular numbers and square numbers.

Problem 7 Examine the following table:

n	1	2	3	4	5	6
T_n	1	3	6	10	15	21
$8T_n$	8	24	48	80	120	168

Find a relationship between $8T_n$ and another geometric number. Justify your relationship in two ways:

(a) by geometric means (draw a figure illustrating the relationship),
(b) by algebraic means.

(D) Trapezoidal Numbers

Trapezoidal numbers are those which can be represented by a (right) trapezoidal array of dots in the manner pictured below. Denote by $\mathcal{T}_n$ the nth trapezoidal number.

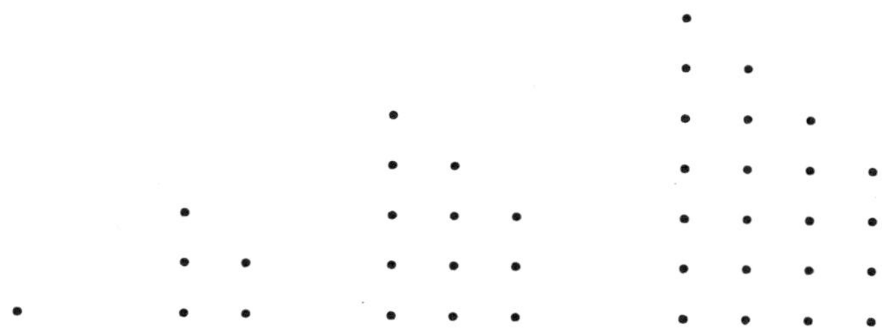

$\mathfrak{T}_1 = 1$ $\mathfrak{T}_2 = 5$ $\mathfrak{T}_3 = 12$ $\mathfrak{T}_4 = 12$

Problem 8 Find $\mathfrak{T}_5$, $\mathfrak{T}_6$, $\mathfrak{T}_7$. Determine a general formula for $\mathfrak{T}_n$ the nth trapezoidal number. Justify your formula.

Problem 9 See if you can make up another type of geometric number and find a gneral formula for it. What are its relationships with S_n, O_n, T_n, and $\mathfrak{T}_n$? Does any partitioning of this number suggest an arithmetical formula?

As you can no doubt see, the possibilities for this topic are endless. Historically the study of these *polygonal numbers* dates back to the time of Pythagoras (about 572-500 B.C.), and in fact they were the principal object of arithmetical investigation through the first century 100 A.D.

As promised, we now turn to some problems involving applications of the formulae just developed, especially the one for T_n.

Problem 10 Find the sum of all integers between 1 and 6,000 (inclusive) which have either 3 or 5 as a factor.

Problem 11

(a) For what (type of) integers n is $3 \mid \mathfrak{I}_n$? Justify.

(b) For what (type of) integers n is $n \mid T_n$? Is $n \mid \mathfrak{I}_n$? Justify.

Problem 12 Which digits can occur as the units digit of oblong numbers? of triangular numbers? Why?

Problem 13 Find (if possible): **(a)** two triangular numbers, **(b)** two oblong numbers which are also square numbers. If no examples exist, explain why.

There are many interesting problems in which you can make use of the formula for T_n found in Problem 4. Four of these are:

Problem 14 For what (type of) integers n is the sum of the first n integers a factor of the product of the first n integers? Justify. For example, $(1 + 2 + 3) \mid (1 \cdot 2 \cdot 3)$, but $(1 + 2 + 3 + 4) \nmid (1 \cdot 2 \cdot 3 \cdot 4)$.

Problem 15 Find all pairs of integers (a, b), such that the sum of all integers greater than a and less than b is 1000.

Problem 16 Given any integer n, let $t(n)$ denote the number of ways n can be written as the sum of two or more consecutive integers.

Examples. $t(5) = 1$ since $5 = 3 + 2$;
$t(8) = 0$ since there is no way 8 can be written as the sum of consecutive integers;
$t(10) = 1$ since $10 = 1 + 2 + 3 + 4$;
$t(15) = 3$ since $15 = 7 + 8 = 4 + 5 + 6 = 1 + 2 + 3 + 4 + 5$.

(a) Find all possible values for $t(p)$ where p is a prime.
(b) Find a general formula for $t(n)$.
Explain your reasoning in each case if you can.

Problem 17 One line divides the plane into 2 regions, two lines (not parallel) divide the plane into 4 regions, three lines (not concurrent) divide the plane into 7 regions. (See figures) How many regions do 4, 5, 6 lines (assume no two lines are parallel, no three lines are concurrent) divide the plane? Find a general formula for the number of regions which n lines divide the plane.

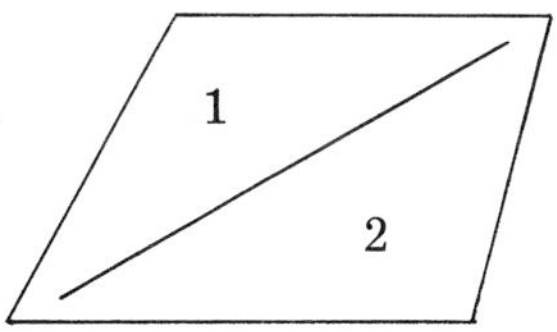

1 line

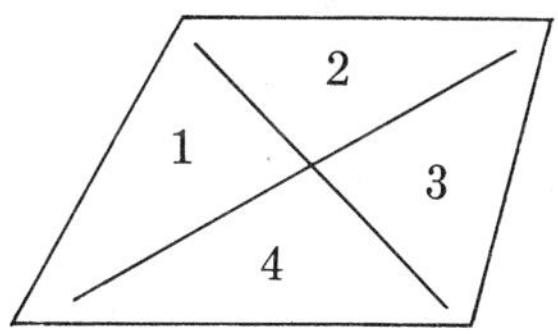

2 lines

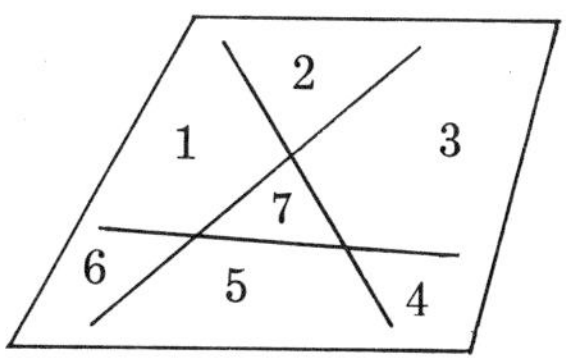

3 lines

Problem 18 Formulate and solve, if you can, a problem which utilizes the formula for T_n .

Closing this chapter is a problem relating triangular numbers to perfect numbers, which were discussed in Chapter 10.

Problem 19 Show that every even perfect number of the form $2^{n-1}(2^p - 1)$ is also a triangular number.

HINTS AND COMMENTS

Problem 1 We see that a 2×2 square can be partitioned into $1 + 3$, a 3×3 square can be partitioned into $1 + 3 + 5$, a 4×4 square can be partitioned into $1 + 3 + 5 + 7$, etc.; each time you add the next successive odd integer, you form the next successive square. Thus the problem is, if you add $1 + 3 + \cdots + 21$, what size square is formed?

Problem 4 How are triangular numbers related to oblong numbers?

Problem 14 Review Problem 8, Chapter 3, page 00.

Problem 16 Review Problem 9, Chapter 2, page 00.

Problem 17 How many regions are added when the fifth line is added? the nth line?

He shall be as a god to me,
who can rightly divide and define.
— *R. Emerson*

12 More on the Division Algorithm

To fulfill the promise made at the end of Chapter 6, we will now take a look at some of the more sophisticated uses of the Division Algorithm. As a bonus you will now get a chance to use some of your algebraic skills, such as squaring a binomial expression [e.g. $(3x + 2)^2$], simplifying expressions, etc.

Recall the statement of the

Division Algorithm For any two integers a and b, with $a \leq b$, there exist integers q and r such that $b = aq + r$ where $0 \leq r \leq a$.

If we take $a = 2$ in the division algorithm, we see that any integer b can be written in the form $b = 2q$ or $b = 2q + 1$, that is, every integer when divided by 2 has a remainder of either 0 or 1; every integer is either even or odd (e.g. $36 = 2 \cdot 18 + 0$ and $47 = 2 \cdot 23 + 1$). Similarly, if $a = 3$, then r is either 0, 1, or 2, so any integer can be written in the form $b = 3q$, $b = 3q + 1$, or $b = 3q + 2$ (e.g. $24 = 3 \cdot 8 + 0$, $37 = 3 \cdot 12 + 1$, $47 = 3 \cdot 15 + 2$).

Problem 1

(a) If $a = 4$ and b is any integer, what forms are possible for b? Give an example of a number representable in each form.

(b) Repeat part (a) for $a = 5$.

(c) Repeat part (a) if a is any fixed integer, say n. That is, $b = nq + r$, so b has the form nq, $nq + 1$, $nq + 2$, etc.

It is not surprising, therefore, that we can discover properties of integers by treating them as algebraic expressions and utilizing some skills of high school algebra, two of which were mentioned above.

Example Prove that every square can be written in the form $4k$ or $4k + 1$, that is, every square when divided by 4 leaves a remainder of 0 or 1 (never 2 or 3!). For example, $36 = 4 \cdot \underline{9}$, $25 = (4 \cdot 6) + 1$.

Proof Every integer can be represented as $2q$ or $2q + 1$. Thus $(2q)^2 = 4q^2 = 4k$, so even squares have the form $4k$; $(2q + 1)^2 = 4q^2 + 4q + 1 = 4(q^2 + q) + 1 = 4k + 1$ for $k = q^2 + q$, so odd squares have the form $4k + 1$.

The idea of this example will be used in the next few problems.

Problem 2 Prove that if x and y are odd integers, then $x^2 + y^2$ cannot be a square.

Problem 3 The example showed that every odd square has the form $4k + 1$. Prove that k is always even, i.e. every odd square can be written in the form $8\ell + 1$ for some integer ℓ. For example, $5^2 = 8 \cdot \underline{3} + 1$, $7^2 = 8 \cdot \underline{6} + 1$.

Problem 4 For what (type of) integers n is $8 \mid (n^2 + 7)$? Prove your answer.

Problem 5 Find all pairs of integers (x, y) for which $2^x = y^2 + 3$. Explain your reasoning.

Problem 6

(a) Find four primes which can be expressed as the sum of two squares.

(b) Find four primes which cannot be expressed as the sum of two squares.

(c) Which primes *cannot* be expressed as the sum of two squares? Explain your reasoning.

Problem 7 A famous theorem of number theory states:

Every integer n can be written as the sum of four or fewer squares.

(a) Illustrate this theorem for $n = 30, 31, 32, 33$.

(b) The statement of the theorem implies that there are some integers which cannot be written as the sum of three or fewer squares. Describe a set of infinitely many such integers. Try to explain why every integer in your set cannot be written as the sum of three or fewer squares. (Experiment with integers greater than 25 before making your guess.)

Problem 8 For which integers n do there exist integers x, y, z such that $n = x^2 + y^2 - z^2$? For example, $3 = 4^2 + 6^2 - 7^2$, $9 = 3^2 + 1^2 - 1^2$.

Problem 9 Prove that every square can be written in the form $3k$ or $3k + 1$ by using the fact that every integer can be written in the form 3ℓ, $3\ell + 1$, or $3\ell + 2$.

Problem 10 Suppose x and y are integers such that $\text{GCF}(3, x) = \text{GCF}(3, y) = 1$. Prove that $x^2 + y^2$ cannot be a square.

Problem 11 For what (type of) integers n is

(a) $5 | (n^2 + 1)$ (b) $5 | (n^2 + 2)$ (c) $5 | (n^2 + 3)$

(d) $5 | (n^2 + 4)$ (e) $5 | (n^2 + 5)$?

Problem 12 What is the largest integer which is a factor of *every* integer of the form $n^5 - n$? Explain your reasoning (i.e. what integer is a factor of every integer in the set $\{2^5 - 2, 3^5 - 3, 4^5 - 4, \ldots\}$?).

Closing this chapter is

Problem 13 In Problem 1(c) we showed that for any integer n, every integer can be written in one of the forms $nq, nq + 1, nq + 2, nq + 3, \ldots, nq + (n - 1)$.

(a) For what (type of) integers n can every integer be represented in one of the forms $nq, nq + 2, nq + 4, \ldots, nq + 2(n - 1)$? For example, if $n = 5$, the statement is true. Every integer can be written in one of the forms; thus $5q, 5q + 2, 5q + 4, 5q + 6, 5q + 8$, since $5q + 6 = (5q + 5) + 1 = 5(q + 1) + 1 = 5k + 1$ for $k = q + 1$, and $5q + 8 = 5k + 3$ using the same reasoning. (We must agree to allow negative values for q in order to obtain the small integers, e.g. we must let $q = -1$ in $5q + 6$ to obtain the integer 1.) For $n = 2$ the statement is false since both $2q$ and $2q + 2$ represent only even integers.

(b) For what (type of) integers n can every integer be written in one of the forms $nq, nq + 3, nq + 6, \ldots, nq + 3(n - 1)$?
In one of the forms $nq, nq + 6, nq + 12, \ldots, nq + 6(n - 1)$?

(c) What is the general situation, i.e. how must the integers n and k related so that every integer can be written in one of the forms $nq, nq + k, nq + 2k, \ldots, nq + k(n - 1)$? What happens otherwise?

HINTS AND COMMENTS

Problem 2 Take any two odd integers x and y. Then any $x = 2k + 1$ for some integer k, any $y = 2\ell + 1$ for another integer ℓ. (In any problem of this sort, to avoid confusion, *be sure to use different letters* for each representation.) Then $x^2 + y^2 = (2k + 1)^2 + (2\ell + 1)^2$. Show that this expression is not a square.

Problem 3 In the example, we had $(2q + 1)^2 = 4(q^2 + q) + 1$, i.e. $k = q^2 + q$. Why is k always even?

Problem 7 The following facts may be useful. By Problem 3, every odd square can be written in the form $8k + 1$. Now, every even square can be written in the form $8k$ or $8k + 4$. (Why?) This theorem was conjectured by the Greek mathematician Diophantus, but it was not completely proved until 1770 by Joseph LaGrange. Thus, you should not be frustrated if you cannot prove all of your conjectures in a short time. But don't give up!

Problem 8 You can use Problem 15, Chapter 3, page 30.

Problem 9 Review the example.

Problem 10 If $\text{GCF}(3, x) = \text{GCF}(3, y) = 1$, what forms are possible for x and y?

Problem 12 The following facts may be useful:
(a) $n^5 - n = n(n^4 - 1) = n(n - 1)(n + 1)(n^2 + 1)$.
(b) $(n - 1)\cdot n\cdot(n + 1)$ is the product of three consecutive integers.

If you choose to represent the various parts in life by holes upon a table, of different shapes—some circular, some triangular, some square, some oblong —and the persons acting these parts by bits of wood of similar shapes, we shall generally find that the triangular person has got into the square hole, the oblong into the triangular, and a square person has squeezed himself into the round hole. . . .

— *S. Smith*

13 Magic Squares

After the "heavy going" of the previous few chapters, the recreational nature of magic squares should come as a welcome relief.

A *magic square* is a square array of integers, all different, such that the sum (or product) of the elements in each row, in each column, and on both diagonals is the same. For example,

5	3	10
11	6	1
2	9	7

is a 3×3 additive magic square since
$5+3+10=11+6+1=$
$2+9+7=5+11+2=$
$3+6+9=10+1+7=$
$5+6+7=10+6+2$, while

20	1	50
25	10	4
2	100	5

is a 3×3 multiplicative magic square.

Problem 1 Complete the following 3×3 additive magic squares.

(a)

16	15	20
		13
14		18

(b)

11		
16		
15		17

(c)

6		
13		
2		

(d)

	7	

(e)

	10	

Problem 2 Construct four 3×3 additive magic squares whose entries are the integers 1, 2, 3, 4, 5, 6, 7, 8, 9. How many different such magic squares are there?

Problem 3 Does there exist a 3×3 additive magic square all of whose entries

(a) are odd integers
(b) are multiples of 5
(c) have 2 as a unit's digit
(d) have a remainder of 3 when divided by 11?

In each case find at least one example or explain why none exist.

Problem 4 In a 3×3 additive magic square, what values are possible for the common sum? What values are possible for the sum of the four corner entries? Explain your reasoning in each case.

Problem 5 Show that in any 3×3 additive magic square, 1 cannot occur in a corner.

Problem 6

$2n - 3$	2	$n + 1$
4	n	$2n - 4$
$n - 1$	$2n - 2$	3

represents a 3×3 additive magic square for any integer $n > 5$. Find at least two other general forms for 3×3 additive magic squares.

Problem 7 Complete the following 4×4 additive magic squares:

(a)

7			19
	10	14	
	9	13	
4			

(b)

			14
		13	
	12		
11			

Problem 8 Construct two 4×4 additive magic squares whose entries are the integers $1, 2, 3, \ldots, 15, 16$.

Problem 9 Construct a 5×5 additive magic square.

Problem 10 Complete the following multiplicative magic squares:

(a)

45	1	75
25	15	

(b)

		98
	14	
2		

(c)

	20	

Problem 11 Does there exist a 3×3 multiplicative magic square all of whose entries are (a) primes? (b) squares? (c) cubes? (d) seventeenth powers? In each case find at least one example or explain why none exist.

Problem 12 In a 3×3 multiplicative magic square, what values are possible for the common product? What values are possible for the product of the four corner entries? Explain your reasoning in each case.

Problem 13 Does there exist a 3×3 magic square which is both additive and multiplicative? (*Note:* The common sum need not be the same as the common product.)

Entries in a magic square need not be restricted to integers, they can be rational numbers. A rational number is one which can be expressed as the quotient of two integers; i.e. a rational number is one which can be expressed in the form a/b where a and b are integers. Rational numbers can be expressed as fractions or decimals. Since we mean "positive integer" when we say "integer," we mean "positive rational number" when we say "rational number." We will discuss rational numbers and decimals in Chapter 15 and 16.

Problem 14 Complete the following additive magic squares. All fractions should be in simplest form.

(a)

		$\frac{5}{6}$
	$\frac{1}{2}$	
$\frac{1}{6}$		

(b)

	$\frac{1}{4}$	

(c)

	.6	

Problem 15 A *magic rectangle* is a rectangular array of integers, all different, such that the sum (or product) of the elements in each row and in each column is the same. Does there exist an additive or multiplicative magic rectangle? Either give a few examples or explain why none exist.

Problem 16 Formulate and solve a problem involving 3×3 magic squares.

A pedagogical comment here is irresistible. In determining all entries in a magic square, one makes many additions and subtractions (or multiplications and divisions). This is much more rewarding than computing 50 independent addition and subtraction problems.

HINTS AND COMMENTS

Problems 4 and 12 How does the common sum (product) depend on the middle entry? Suppose the following is a magic square:

a	b	c
d	e	f
g	h	i

Let S denote the common sum. Then

$$\begin{aligned} S &= a + e + i \\ S &= d + e + f \\ S &= g + e + c \\ \hline \end{aligned}$$

Add these equations and see what happens.

Problems 8 and 9 There are 880 different 4×4 additive magic squares with entries $1, 2, 3, \ldots, 15, 16$; while there are 60,000 different 5×5 additive magic squares with entries $1, 2, 3, \ldots, 24, 25$.

Problem 14(a) Look closely at the example. After completing this problem, formulate a general method for constructing 3×3 additive magic squares whose entries are rational numbers.

14 More on Primes

> But it's just a crime when you're past your prime.
>
> — *Daisie Mae*

In this chapter we discuss more properties of primes, specifically: (1) how many primes are there? and (2) how are the primes distributed among all the integers?

In Chapter 2 we learned a method (somewhat tedious for large integers) for determining whether a given integer is a prime. "How long must we do this?", "Is there a largest prime?" or in other words "Are there infinitely many primes?"

Problem 1 Let $f(n)$ denote the number of primes less than or equal to n. For example, $f(7) = 4$ because 2, 3, 5, 7 are all the primes less than or equal to 7, and $f(20) = 8$ (what are the primes?).

(a) Compute $f(25)$, $f(50)$, $f(97)$, $f(98)$, $f(99)$, $f(100)$, $f(101)$.

(b) For what (type of) integers n is

$$\frac{f(n-1)}{n-1} < \frac{f(n)}{n} \ ?$$

Explain your reasoning.

Euclid [300 B.C. (?)], the famous Greek mathematician, most noted for his development of plane geometry, proved that there are, indeed, infinitely many primes. His proof is a classic example of indirect reasoning, and for that reason is presented here.

Euclid's Theorem *There are infinitely many primes.*

Discussion and Proof: We prove that there are infinitely many primes by showing there cannot be a finite number of them. (This type of thinking is the essence of indirect reasoning, which we first discussed in Chapter 2.) To be specific, only two possibilities exist:

(1) There are an infinite number of primes.

(2) There are a finite number of primes.

We prove that statement (1) is true by assuming that statement (2) is true and deriving a contradiction. So suppose (2) is true; suppose there are only a finite number of primes, $2, 3, 5, \ldots, P$ where P denotes the largest prime.

The strategy of the proof is the following: We have only a finite number of primes $2, 3, 5, 7, \ldots, P$, and we also know by the Fundamental Theorem that every integer can be written as a product of these primes. The key step now is to find an integer which cannot be written as the product of these primes that will contradict our supposition that $2, 3, 5, \ldots, P$ are the only primes. Consider the integer

$$K = 2 \cdot 3 \cdot 5 \cdots P + 1 .$$

By the Fundamental Theorem, either K is a prime or K can be written as the product of primes. Well, K is not a prime since $K > P$ and P is the largest prime, so K must have a prime factor, call it q. Now, $q \mid 2 \cdot 3 \cdot 5 \cdots P$ since this product contains all the primes, so q must be a factor of the difference of these two integers, i.e. $q \mid K - (2 \cdot 3 \cdot 5 \cdots P)$. But this difference $K - (2 \cdot 3 \cdot 5 \cdots P) = 1$. Alas $q \mid 1$! An absurdity!

Thus, since the supposition that there are only a finite number of primes led to the conclusion that one of them was a factor of 1, the supposition is false. So statement (2) cannot be true—leaving only the possibility that (1) is true: There are, indeed, infinitely many primes.

Graphically, the idea behind the proof is that there is another prime not equal to $2, 3, 5, \ldots, P$ in the interval from P to $K = 2 \cdot 3 \cdot 5 \cdots P + 1$.

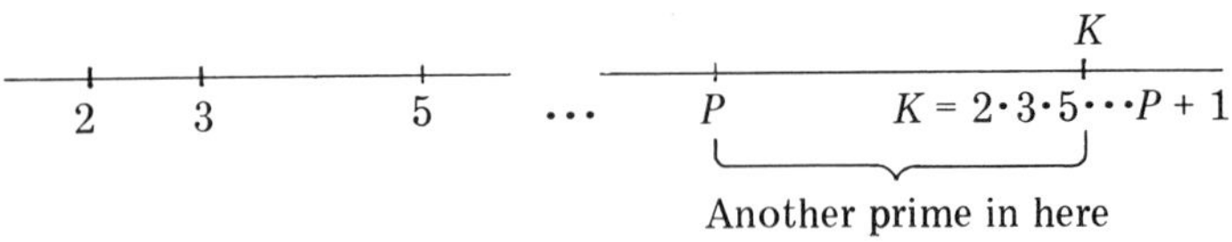

Example $(2 \cdot 3 \cdot 5) + 1 = 31$ is a prime not equal to 2, 3, 5, while $(2 \cdot 3 \cdot 5 \cdot 7 \cdot 11 \cdot 13) + 1 = 30{,}031 = 59 \cdot 509$ which contains a prime factor not equal to 2, 3, 5, 7, 11, or 13.

Problem 2 Does Euclid's proof work if we take
(a) $K = 2 \cdot 3 \cdot 5 \cdots P - 1$?
(b) $K = 2 \cdot 3 \cdot 5 \cdots P + 6$?
(c) $K = P! + 1$?
Explain your reasoning in each case.

Problem 3 For any integer $n > 2$, prove that there is at least one prime between n and $n!$.

Problem 4 How many primes are there between the integers $2 \cdot 3 \cdot 5 \cdot 7 \cdots p$ and $(2 \cdot 3 \cdot 5 \cdot 7 \cdots p) + p$? Justify your answer.

Problem 5 Let p_n denote the nth prime $[p_1 = 2, p_2 = 3, p_3 = 5$, etc.]. Compute $(p_1 p_2 p_3 \cdots p_n) + 1$ for $n = 3, 4, 5, 6$. Can the integer $p_1 p_2 p_3 \cdots p_n + 1$ be
(a) a square (b) the sum of two squares
for any value of n? Justify.

Problem 6 The interval of the six consecutive integers 2, 3, 4, 5, 6, 7 contains four primes. Are there any other intervals of six consecutive integers containing four primes? three primes? Why?

Problem 7

(a) Find two values of n for which

(1) $n + 2,\ n + 3,\ n + 4$ are all composite.

(2) $n + 2,\ n + 3,\ n + 4\ n + 5$ are all composite.

(b) Show that $8! + 2,\ 8! + 3,\ 8! + 4,\ 8! + 5,\ 8! + 6,\ 8! + 7,\ 8! + 8$ are all composite.

(c) Describe a sequence of 1000 consecutive composite integers.

(d) Is there any limit to the number of consecutive composite integers, i.e. for any integer k does there exist a sequence of k consecutive composite integers? Why?

Problem 7 shows that the primes are "spread out pretty thin"; however, they also "cluster" in groups of two.

Problem 8 A *twin prime* is a set of two consecutive odd primes, e.g. (3, 5), (11, 13), (59, 61). Find all twin primes $(p, p + 2)$ with $p \leq 101$.

Problem 9 Let p and q be consecutive odd primes. Prove that $(p + q)$ has at least three (not necessarily distinct) prime factors. For example: $7 + 11 = 2 \cdot 3 \cdot 3$, $3 + 5 = 2 \cdot 2 \cdot 2$, $13 + 17 = 2 \cdot 3 \cdot 5$. If (p, q) is a twin prime, which primes are always factors of $(p + q)$? Why?

Problem 10 Find all integers n for which $(2^n - 1,\ 2^n + 1)$ is a twin prime.

Problem 11 Find all primes p for which $(p^2 + 4,\ p^2 + 6)$ is a twin prime. Explain your reasoning.

There are infinitely many primes, but are there infinitely many twin primes? No one knows. This problem has remained unsolved for centuries.

Analogous to twin primes are prime triplets which form the content of

Problem 12 A *prime triplet* is a set of three primes (p, q, r) such that $q - p = r - q$. How many "twin" prime triplets are there, i.e. set of primes (p, q, r) with $q - p = r - q = 2$? Find all prime triplets (p, q, r) with $p = 3, 5, 7, 11$ and $r < 100$. Consult your list of primes for Chapter 2, Problem 1. Can you make any conjectures about the number $q - p$ for $p \geq 5$? Justify your conjectures if you can.

Problem 13 The concept of twin prime and "twin" prime triplet can be generalized to allow the gap between the primes to be any even integer. Find all pairs of primes (p, q) with

(a) $q - p = 4$ (b) $q - p = 6$ for $p \leq 101$.

Problem 14

(a) Find all prime triplets (p, q, r) with $r - q = q - p = 4$.

(b) Find five prime triplets (p, q, r) with $r - q = q - p = 6$. There appears to be infinitely many such prime triplets. Explain why this is plausible (i.e. where does the type of reasoning used for Problem 12 and Problem 14(a) "break down"?).

Problem 15 Formulate and solve a problem about triplets of primes (p, q, r) with $r - q = q - p = 2k$ for some integer k.

HINTS AND COMMENTS

Problem 3 Examine $n! - 1$.

Problem 4 $2 \mid (2 \cdot 3 \cdot 5 \cdot 7 \cdots p) + 2, \quad 3 \mid (2 \cdot 3 \cdot 5 \cdots p) + 3$, etc.

Problem 5(a) You've done this problem twice before in another context—see Chapter 3. For 5(b), review Chapter 12.

Problem 7 Euclid's Theorem and Problem 7 bring out two facts about prime numbers:

(1) There are infinitely many primes.
(2) There are arbitrarily large gaps in the sequence of primes.

Are these two statements contradictory? See if you can convince your classmate that, indeed, they are not.

Problem 9(b) In each case you are dealing with three *consecutive* integers.

Problem 12 Use the Division Algorithm with $a = 6$.

Problem 14(a) Use the Division Algorithm with (a) $a = 12$, (b) $a = 18$, or some other suitable value for a.

15 Rational Numbers

> Man is a rational animal who always loses his temper when he is called upon to act in accordance with the dictates of reason.
>
> — *O. Wilde*

Thus far we have restricted ourselves to problems involving integers, but there are, as I'm sure you are aware, other types of numbers—namely rational numbers and irrational numbers, which together form what is commonly called the "real numbers." (If you've had any advanced algebra you have also encountered imaginary and complex numbers—an area we will not discuss.)

A *rational number** is one which can be expressed as the quotient of two integers, i.e. a rational number is one which can be expressed in the form a/b where a and b are integers. Hence, according to the definition, all integers n are rational since $n = n/1$; i.e. the set of integers is a subset of the set of rational numbers.

Any given rational number can be expressed in infinitely many different ways as a quotient of integers, e.g. $\frac{2}{3} = \frac{4}{6} = \frac{6}{9} = \frac{8}{12} = \cdots$. Most of the time, however, we will avoid this ambiguity by taking the given rational number to be in *simplest form,* i.e. in the form a/b where $\text{GCF}(a, b) = 1$. (In our example $2/3$ is in simplest form.)

Problem 1 Find the simplest form of each of the following rational numbers:

(a) $\frac{1470}{3690}$ (b) $\frac{4171}{6407}$ (c) $\frac{3137}{7583}$

*We will maintain our convention that *rational number* means *positive rational number.*

Problem 2 For what (type of) integers n is the rational number $(n + 3)/(n^2 + 2n)$ in simplest form? Justify your answer.

Problem 3 If $\frac{a}{b}$ and $\frac{c}{d}$ are two rational numbers in simplest form, does it follow that

(a) $\dfrac{ad + bc}{bd}$ (b) $\dfrac{ac}{bd}$ (c) $\dfrac{ad}{bc}$ is in simplest form?

In each case either prove that the rational number is always in simplest form or give three examples where it fails to be in simplest form.

Problem 4 A rational number a/b is said to be *proper* if $a < b$, that is $a/b < 1$. How many proper rational numbers in simplest form are there whose denominator is

(a) 97 (b) 91 (c) 99?

Again, use your brain instead of your pencil.

Problem 5 Show that in the following sets of rational numbers, an even number of them are in simplest form:

$$S = \left\{\frac{1}{n}, \frac{2}{n}, \frac{3}{n}, \ldots, \frac{n-2}{n}, \frac{n-1}{n}\right\}.$$

Try some examples first—take $n = 5, 12, 16$.

A good part of your early mathematical training probably consisted of learning how to add, subtract, multiply, and divide rational numbers. I now give you a chance to review these techniques while solving some problems.

Problem 6 A rational number is called *unitary* if its numerator is 1, i.e. if it has the form $1/n$. Can the sum of two different unitary rational numbers be a unitary rational number? Either give three examples or explain why none exist.

Problem 7 Can the sum of three unitary rational numbers equal 1? Either find all ways this can be done (this will involve some general reasoning) or explain why it can never be done.

Problem 8

(a) Find three distinct integers k_1, k_2, k_3, such that the sum of the three unitary rational numbers $\frac{1}{k_1} + \frac{1}{k_2} + \frac{1}{k_3}$ is less than $\frac{1}{2}$ but as close to $\frac{1}{2}$ as possible.

(b) Now find four distinct integers k_1, k_2, k_3, k_4 such that the sum of the four unitary rational numbers $\frac{1}{k_1} + \frac{1}{k_2} + \frac{1}{k_3} + \frac{1}{k_4}$ is less than $\frac{1}{2}$ but as close to $\frac{1}{2}$ as possible.

(c) Do the same for 5.

(d) What is the general strategy?

Problem 9 Compute:

(a) $\frac{1}{1 \cdot 2} + \frac{1}{2 \cdot 3} + \frac{1}{3 \cdot 4}$

(b) $\frac{1}{1 \cdot 2} + \frac{1}{2 \cdot 3} + \frac{1}{3 \cdot 4} + \frac{1}{4 \cdot 5} + \frac{1}{5 \cdot 6} + \frac{1}{6 \cdot 7} + \frac{1}{7 \cdot 8}$

(c) $\frac{1}{1 \cdot 2} + \frac{1}{2 \cdot 3} + \cdots + \frac{1}{9 \cdot 10}$

(d) $\frac{1}{1 \cdot 2} + \frac{1}{2 \cdot 3} + \cdots + \frac{1}{19 \cdot 20}$

Find and prove, if you can, a general formula for the sum

$$\frac{1}{1 \cdot 2} + \frac{1}{2 \cdot 3} + \frac{1}{3 \cdot 4} + \cdots + \frac{1}{(n-1) \cdot n}\ .$$

Problem 10 For which integers n is

$$\frac{1}{2^2} + \frac{1}{3^2} + \frac{1}{4^2} + \cdots + \frac{1}{n^2} < 1? \quad \text{Justify.}$$

Problem 11 In how many ways can $\frac{1}{2}$ be written as the sum of two unitary rational numbers? Answer the same question for $\frac{1}{3}, \frac{1}{6}, \frac{1}{8}$. Find a general formula (in terms of n) which tells you how many ways $\frac{1}{n}$ can be written as the sum of two unitary rational numbers. Prove your answer if you can.

Problem 12 Find the smallest rational number which when divided by $\frac{6}{35}$, $\frac{10}{21}$, and $\frac{15}{49}$ yields an integer as the quotient in each case.

In the next two problems you will investigate two common errors made in algebra and/or arithmetic. Instead of simply saying they are wrong, we are going to try to see if they are ever correct.

Problem 13 One error is $\frac{2}{3} + \frac{3}{5} = \frac{2+3}{3+5} = \frac{5}{8}$. Is this phenomenon ever true, i.e. do there exist integers a, b, c, d such that $\frac{a}{b} + \frac{c}{d} = \frac{a+c}{b+d}$? Either show that no such integers exist or determine how these integers a, b, c, d must be related in order for this phenomenon to occur. Give three examples.

Problem 14 Another common error is $\frac{5+\cancel{4}\cdot\cancel{3}}{6+\cancel{4}\cdot\cancel{7}} = \frac{5+3}{6+7} = \frac{8}{13}$. Is this phenomenon ever true, i.e. do there exist integers a, b, c, d, e such that $\frac{a+e\cdot b}{c+e\cdot d} = \frac{a+b}{c+d}$? Either show that no such integers can exist or determine how the integers a, b, c, d must be related in order for this phenomenon to occur. Give three examples.

The concept of order was not discussed for integers because it is easy to tell which of two integers is the larger. For rational numbers it is not quite so obvious, so

Problem 15

(a) Arrange the following rational numbers in increasing order: $\frac{143}{210}$, $\frac{47}{70}$, $\frac{2}{3}$, $\frac{24}{35}$, $\frac{71}{105}$.

(b) If $\frac{a}{b}$ and $\frac{c}{d}$ are rational numbers with $\frac{1}{2} \leq \frac{a}{b} \leq \frac{c}{d} < 1$, arrange the following in increasing order: $\frac{b}{a}$, $\frac{d}{c}$, $\frac{bd}{ac}$, $\frac{b+d}{a+c}$, 1.

Problem 16

(a) Find six rational numbers between $\frac{1}{3}$ and $\frac{2}{5}$.

(b) In general, prove that between any two rational numbers there are infinitely many other rational numbers.

All integers are rational numbers. In these last few problems you will be asked to determine when, if ever, certain rational expressions are integers.

Problem 17

(a) For what values of n is

(i) $\dfrac{n^2 - 5n + 6}{n - 2}$ (ii) $\dfrac{n^3 - 9n^2 + 26n - 24}{n - 2}$

an integer?

(b) For what pairs of integers (m, n), where $\mathrm{GCF}(m, n) = 1$, is $\dfrac{m^2 + n^2}{m + n}$ an integer?

Justify your answers in each case.

Problem 18 For what (type of) integers n is

(a) $\dfrac{n^2}{3} + \dfrac{n}{2}$ (b) $\dfrac{n^3}{3} + \dfrac{n^2}{2} + \dfrac{n}{6}$ an integer?

Explain your reasoning in each case.

Problem 19

(a) Can the sum of a rational number (not equal to 1) and its reciprocal ever be an integer, i.e. do there exist integers a and b such that $\dfrac{a}{b} + \dfrac{b}{a}$ is an integer? Either find three examples or explain why none exist.

(b) Let p_n denote the nth prime. For which integers n is $\dfrac{1}{p_1} + \dfrac{1}{p_2} + \dfrac{1}{p_3} + \cdots + \dfrac{1}{p_n}$ an integer? Justify.

HINTS AND COMMENTS

Problems 1 and 2 You can use the Euclidean Algorithm.

Problem 5 Look at $\frac{k}{n}$ and $1-\frac{k}{n}$.

Problem 9 For any integer k, $\frac{1}{(k-1)k} = \frac{1}{k-1} - \frac{1}{k}$, e.g. $\frac{1}{2\cdot 3} = \frac{1}{2} - \frac{1}{3}$.

Problem 11 Suppose that $\frac{1}{2} = \frac{1}{m} + \frac{1}{k}$.
Then $mk = 2m + 2k$, or $m = \frac{2k}{k-2}$.
The question now is: For what values of k is m an integer? The general problem is more difficult. If $\frac{1}{m} + \frac{1}{k} = \frac{1}{n}$, then $nk + mn = mk$, or $mk - nk - mn = 0$. Here is a trick. (I dislike using tricks like this, but I will do it anyway because I feel the problem is so interesting that if you solve it with this help, you will still get something out of the problem.) Now, $mn - mk - nk = 0$, so $mk - nk - mn + n^2 = n^2$, or $(m-n)(k-n) = n^2$. Why did I add n^2 to both sides of the equation? Good question! I did it because I wanted to factor the left-hand side of the equation so as to make use of my knowledge of factors of an integer. Adding n^2 to the left-hand side allows me to factor it, and if I add n^2 to the left-hand side, I must also add it to the right-hand side to maintain the equality.

You try to finish the problem.

Problems 13 and 14 Write the problem in terms of an algebraic equation, and then simplify the equation.

Problem 19 Use indirect reasoning, i.e. suppose $\frac{a}{b} + \frac{b}{a} = n$, n an integer. Then $a^2 + b^2 = nab$. Is this equation possible?

16 Irrational Numbers

> Irrationally held truths may be more harmful than reasoned errors.
>
> — *B. Huxley*

There is no simple definition of an irrational number—it is best thought of as a number which is not rational. The discovery that not all numbers were rational was a traumatic experience for the early Greeks. They were astonished to find that rational numbers (which, you recall, includes the integers) could not account for even the simplest geometric notions. Their classical example was that the length of the diagonal of a square whose side was 1 unit long was not rational, that is:

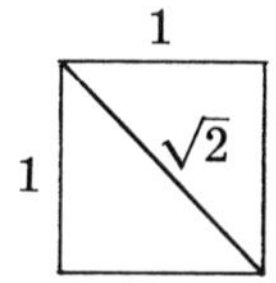

Problem 1 Show that $\sqrt{2}$ is irrational (not rational) using the Fundamental Theorem of Arithmetic. (See the Hints and Comments if you need help.)

To reinforce the technique of Problem 1, try:

Problem 2 Show that $\sqrt{3}$ and $\sqrt{6}$ are irrational numbers.

Problem 3 Show that $1 + \sqrt{2}$ and $3 - \sqrt{6}$ are irrational numbers.

Problem 4 Show that $\sqrt[3]{5}$ is an irrational number.

Problem 5 Show $\sqrt{2}+\sqrt{3}$ is an irrational number.

Problem 6 Can you formulate a strategy for numbers involving square roots, cube roots, etc. to be irrational? Illustrate your strategy by proving that $\sqrt[n]{n}$ is irrational for $n \geq 2$.

Probably the earliest known irrational number was the so-called "golden mean." One example illustrating its definition is the following: Consider the line segment AC divided into two segments by point B. (See Fig. 1.) Then a denotes the length of AB, b denotes the length of BC. Suppose the point B divides this segment in such a way that $\frac{a+b}{a} = \frac{a}{b}$. This common ratio $\frac{a+b}{a} = \frac{a}{b}$ is called the "golden ratio" or "golden mean," and the ratio of the line segments is called the "golden section."

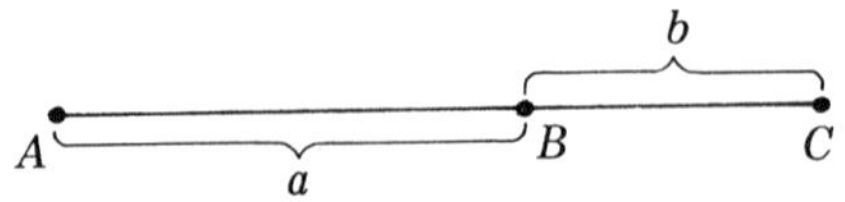

Figure 1

Problem 7 Show that the golden ratio is, indeed, an irrational number. Keep in mind that you know nothing about what type of numbers a and b are, you only know that $\frac{a+b}{a} = \frac{a}{b}$.

Many early civilizations recognized the esthetic and artistic properties of geometric figures which exhibited the "golden mean." (It is for this reason that I referred to it earlier as "probably the first known irrational number.")

Many people have said that the Greeks constructed the floors of their temples in a rectangular fashion (called the "golden rectangle") such that $\frac{a+b}{a} = \frac{a}{b}$. (See Fig. 2.)

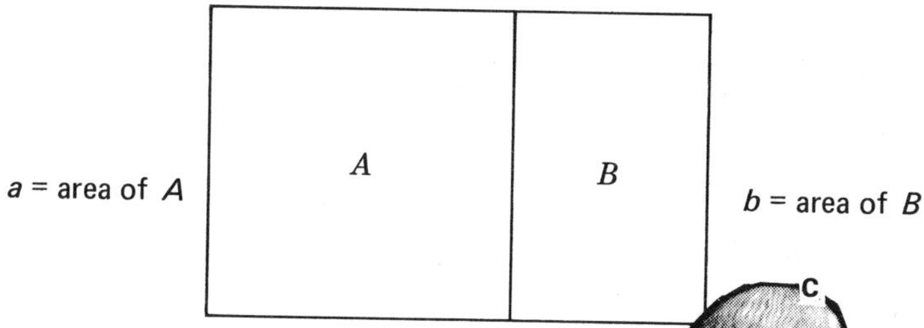

Figure 2

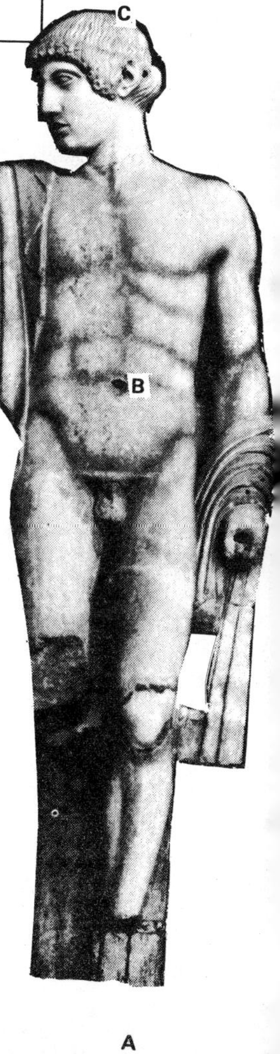

Another (sometimes doubted) statement tells us that the Egyptians built their pyramids in such a way that for each triangular face, $\frac{s}{h} = \frac{h}{b}$. (See Fig. 3.)

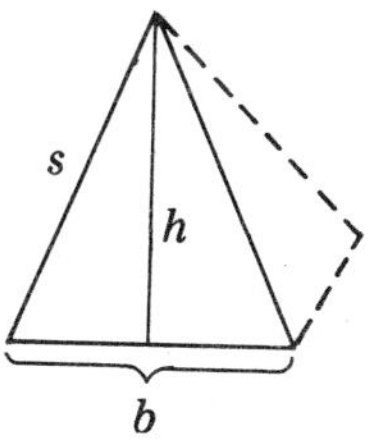

Figure 3

And, finally, a "perfect" man was considered by some ancient Greeks to be one proportioned so that $\frac{a+b}{a} = \frac{a}{b}$. (See Fig. 4.)

Figure 4

An example of the use of the golden ratio today are file cards in either of the two standard sizes 3×5 or 5×8, which are close approximations to the "golden rectangle."

At this point you should feel that most numbers of the form $\sqrt[m]{n}$, $a + \sqrt[m]{n}$, numbers involving radicals, etc., are irrational numbers. There are other irrational numbers (π, for example), but my purpose here is *not* to give a complete exposition of irrational numbers but rather to indicate that such numbers do exist.

You probably know that the sum, product, and quotient of any two (non-zero) rational numbers is again a rational number, but is the same true for irrational numbers? We have seen in Problem 5, for instance, that $\sqrt{2} + \sqrt{3}$ is irrational.

Problem 8

(a) Is the sum of any two irrational numbers always irrational? Why?

(b) How about the product? the quotient? Why?

Problem 9 We saw in Problem 3 that $1 + \sqrt{2}$ and $3 - \sqrt{6}$ were irrational. Is the sum of a rational number and an irrational number always irrational? Why?

Problem 10 Can $a + \sqrt{b} = c + \sqrt{d}$ if $a \neq c$ and $b \neq d$? Justify.

Problem 11 Which is larger:

(a) $\sqrt{6} + \sqrt{7}$ or $\sqrt{5} + \sqrt{8}$?

(b) $\sqrt{5} + \sqrt{8}$ or $\sqrt{3} + \sqrt{11}$?

Justify your answer in each case without using tables.

Problem 12 A common error involving square roots of numbers is to assume that $\sqrt{a+b} = \sqrt{a} + \sqrt{b}$ where a and b are integers.

(a) Give three examples where this statement is false.

(b) Are there any values of a and b for which it is true? Justify.

The most common way of distinguishing between rational and irrational numbers is to look at their decimal expansions. For example:

$$\frac{1}{4} = .25 \qquad \sqrt{3} = 1.732\ldots$$

$$\frac{1}{6} = .16666\ldots \qquad \pi = 3.14159\ldots$$

The criterion is:

A number is rational if its decimal expansion is either (1) finite (like $\frac{1}{4}$) or (2) repeating after some point (like $\frac{1}{6}$).

This statement will not be proved in general because, I feel, the excessive notation needed will obscure the point. Looking at the next three examples should, however, make the statement believable.

EXAMPLES

(1) $n = .2345$ is rational; $.2345 = \frac{2324}{10{,}000}$, so it can be written as the quotient of two integers.

(2) $n = .123123123\ldots$ is rational. We see that the decimal expansion repeats every three digits, so that $1000n = 123.123123\ldots$ has the same "decimal part" as n.

We have $1000n = 123.123123\ldots$

and $\underline{\quad n = \quad .123123\ldots}$

so $999n = 123$,

or $n = \frac{123}{999} = \frac{41}{333}$, which is rational.

Suppose the decimal expansion of a given number does not repeat right away. There is really no difference in showing such numbers to be rational.

(3) $n = .23474747\ldots$. Here the decimal expansion repeats every two digits, so look at $100n$:

$$\begin{aligned} 100n &= 23.474747\ldots \\ n &= .234747\ldots \\ \hline 99n &= 23.24 \end{aligned}$$

Hence $99n$ is rational since it has a finite decimal expansion, and thus so is $\frac{99n}{99} = n$. In fact,

$$n = \frac{2324}{100 \cdot 99} = \frac{2324}{9900} = \frac{581}{2475}.$$

Problem 13 Find the rational number represented by each of the following decimals:

(a) .24786

(b) .373737 . . .

(c) 2.14232323 . . .

(d) 3.123456456456 . . .

Problem 14 Is .123456789101112131415 . . . rational? Explain.

Problem 15 For what (type of) integers n does $1/n$ have a finite decimal expansion?

Problem 16 Using decimal expansions, prove that there is an irrational number between any two rational numbers. For example, $1.41 < \sqrt{2} < 1.42$ (why?).

Problem 17 Suppose $\alpha = .127\ldots$

$\beta = .126\ldots$

i.e. you are given the first three digits of the decimal expansions of two numbers.

(a) If α and β irrational, can $\alpha = \beta$? That is, can you choose the remaining digits in the decimal expansions of α and β? Explain your reasoning.

(b) If α and β are rational, can $\alpha = \beta$? Explain.

HINTS AND COMMENTS

Problem 1 This problem offers another chance to use indirect reasoning. Suppose $\sqrt{2}$ is rational, i.e. $\sqrt{2} = m/n$, where m, n are integers. Then $2 = m^2/n^2$, or $2n^2 = m^2$. Now, count the number of times 2 could appear as a factor on both sides of this equation, to derive your contradiction; i.e. 2 must appear an odd number of times on the left, while 2 must appear an even number of times on the right. Why?

Problem 7 If $\dfrac{a+b}{a} = \dfrac{a}{b}$, then $a^2 = ab + b^2$. Divide both sides of this equation by b^2. Can $\dfrac{a}{b}$ now be rational?

Problems 10, 11, and 12 You want to remove the radical signs—how can you do this?

17 Probability

> The laws of probability, so true in general, so fallacious in particular.
>
> — *E. Gibbon*

> Probability must atone for want of truth.
>
> — *M. Prior*

You have undoubtedly seen the words "probability" and "probable" used many times. For example:

(1) "There is probably life on another planet."

(2) "The Minnesota Vikings are probably the best football team in the country."

(3) "There is a 30% probability of rain tomorrow."

(4) "There is a 95% probability of surviving an attack of pneumonia if sulfa drugs are used."

As these examples indicate, the word "probability" is subject to a wide range of meanings. Sentences (1) and (2) are examples of a philosophical interpretation of the word "probability," while sentences (3) and (4) are examples of a mathematical interpretation. Philosophically, probability is a state of mind—none of our knowledge is certain, the degree of our belief in any proposition is its probability. You might try to find some other examples of "probability" used in this way.

But this is, after all, a mathematics book, so we shall be concerned with the mathematical view of probability as expressed in statements (3) and (4). In the mathematical sense, probability is the statistical concept of relative frequency (based usually on past experience). For instance, statement (4) was made because after checking the records of many pneumonia patients who had sulfa drugs administered to them, it was found that 95% survived—i.e. the relative frequency of survivors was .95. Statement (3) was also based on past experience—of all days with similar temperature, humidity, cloud conditions, etc., it rained on .3 of them.

Thus (mathematical) probability stems from the idea of the relative frequency of one (or more) event(s) in a set of events, the relative frequency of one (or more) outcome(s) in a situation which can have several different outcomes. If you draw a card from a standard deck, 52 outcomes are possible; if you choose a digit at random, 10 outcomes are possible; if you toss a true die, 6 outcomes are possible. The set of all possible outcomes for any situation is called a *sample space,* denoted S. For example, the sample space for choosing a digit is $S = \{0, 1, 2, 3, 4, 5, 6, 7, 8, 9\}$. The simplest situation to analyze is one in which all outcomes are considered to be equally likely; we assume each face of the true die has the same chance to appear; each card of the standard deck has the same chance to be drawn. We will assume, therefore, that for all our problems, all outcomes in the sample space have the same likelihood of occurring.

Let's consider an example.

Example Choose a digit at random. The sample space is $S = \{0, 1, 2, 3, 4, 5, 6, 7, 8, 9\}$. Find the probability that this digit is

(a) even **(b)** odd **(c)** a prime **(d)** a square **(e)** a multiple of 3.

For each of these situations there are outcomes; in (a) there are 5 favorable outcomes $\{0, 2, 4, 6, 8\}$, so the probability in (a) is $5/10 = 1/2$. Similarly, in (b), (c), (d) and (e) the probabilities are 1/2, 2/5, 3/10, and 2/5 (considering 0 to be a multiple of 3) respectively. To reiterate, the relative frequency of even digits in the sample space of all digits is 1/2, of primes is 2/5, of squares is 3/10 (0 is not considered a square), and of multiples of 3 is 2/5.

This example indicates the fundamental formula for probability (when all outcomes in the sample space are equally likely):

$$\text{Probability of the occurrence of an event} = \frac{\text{Number of favorable outcomes}}{\text{Total number of outcomes in the sample space}}$$

In the example, the event—the chosen digit is even—has five favorable outcomes among the ten possible outcomes in the sample space, and hence a probability 1/2 of occurring.

In keeping with the spirit of this book, all of our problems will deal with numbers instead of the customary coins, dice, or cards.

Problem 1 What can you say about an event if it has probability
(a) 0 (b) 1 of occurring?

Problem 2 Let $S = \{1, 2, 3, 4, \ldots, 99, 100\}$. Choose an integer n at random from S. Find the probability that:
(a) $n = 73$ (b) n is even (c) n is odd (d) n is prime (e) n is composite (f) $n(100 - n) > 1000$ (g) n is a square (h) n is a cube (i) n is either a square or a cube (j) n is a multiple of 5 (k) n is a multiple of 3 (l) n is either a multiple of either 2, 3, or 5.

Problem 3 Let $S = \{1, 2, 3, \ldots, 5,999, 6,000\}$. Choose an integer n at random from S. Find the probability that:

(a) n has a remainder of 6 when divided by 11.
(b) n has a remainder of 5 when divided by 11.
(c) n is relatively prime to 6 (See Chapter 7 for a definition of "relatively prime" in case you forgot.)
(d) n^2 has 1 as its unit's digit.
(e) 2^n has 2 as its unit's digit.
(f) 2^n has 12 as its last two digits.

Problem 4 Again let $S = \{1, 2, 3, \ldots, 99, 100\}$. This time choose two distinct integers n_1 and n_2 at random from S. Find the probability that:

(a) $n_1 + n_2 > 195$ **(b)** $n_1 + n_2 > 199$

(c) $n_1 + n_2 > 2$

(d) $n_1 + n_2$ is even **(e)** $n_1 + n_2$ is odd

(f) $n_1 + 2n_2$ is even **(g)** $n_1 + 2n_2$ is odd

(h) $n_1 + n_2$ is a multiple of 5

Several of the preceding problems have hinted at the fundamental difficulty in computing probabilities—that of counting the number of elements in the set of favorable outcomes and in the sample space. (Who would have thought that counting could be so difficult?) In the remaining problems you may need to invent some ways of counting the elements in these sets. If you have trouble, there is some help in the Hints and Comments.

Problem 5 Changing notation somewhat let D_k denote set of all integers with k digits (e.g. D_4 is the set of all 4-digit integers, etc.). Choose an integer n at random from

(a) D_4 **(b)** D_5 **(c)** D_8 **(d)** D_{11}.

In each case find the probability that n has distinct digits, i.e. that no digit occurs twice in the integer.

Problem 6 Let $S = \{1, 2, 3, 4, \ldots, 999{,}999\}$. Choose an integer n at random from S. Find the probability that:

(a) n has exactly five 7's among its digits

(b) n has no 7's among its digits

(c) n is a palindrome.

Problem 7 Let S be the set of all 3-digit integers formed by using the digits 1, 2, 3, 4, 5, each only once. Choose an integer n at random from S. Find the probability that $9 \mid n$.

Problem 8 Let S be the set of all 6-digit integers formed by using the digits 1 and 2, using each digit at least once. Choose an integer n at random from S. Find the probability that
(a) $4 \mid n$ (b) $8 \mid n$ (c) $3 \mid n$.

Problem 9 Let S be the set of all 9-digit integers formed by using each of the digits 1, 2, 3, 4, 5, 6, 7, 8, 9 only once. Choose an integer n at random from S. Find the probability that $11 \mid n$.

Problem 10 This problem differs from the first nine because the sample space is infinite (but you can still work the problem if you try). Let S be the set of all positive integers. Choose any four integers a, b, c, d at random from S. Find the probability that $ab - cd$ (which may be negative) is odd.

I close the chapter with two problems which may seem slightly out of place: however, they will give you a chance to review two basic ideas from your high school algebra course in a nonroutine way. A little review never hurt anyone.

Problem 11 Let S be the set of all systems of equations of the form

$$\begin{aligned} ax + by &= 10 \\ x + 2y &= 15, \end{aligned}$$

where a and b are integers with $1 \leq a \leq 20$ and $1 \leq b \leq 20$. Choose one of these systems of equations at random from S. Find the probability that this system has a solution (a solution set may have rational or negative numbers).

Problem 12 Let S be the set of all quadratic equations of the form $x^2 + bx + c = 0$ where b and c are integers with $1 \leq b \leq 10$ and $1 \leq c \leq 10$. Choose one of the quadratic equations at random from S. Find the probability that this equation has real solutions.

HINTS AND COMMENTS

For each of these problems first describe the sample space, and then determine the number of elements in the sample space and the number of favorable outcomes.

Problem 2 (i) and (l) See Chapter 1, Problems 7 and 8.

Problem 4 You can think of counting the number of elements of the sample space as filling two squares:

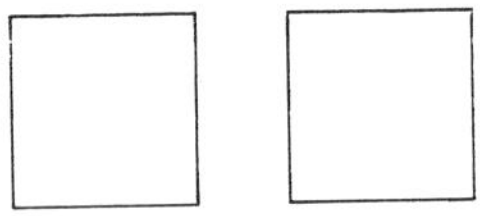

the first with n_1, the second with n_2. The first square can be filled any of 100 ways: the second can be filled in 99 ways since there are 99 integers remaining. Thus the sample space contains 9900 elements.

(d) and **(e)** To find the number of pairs n_1 and n_2 such that $n_1 + n_2$ is odd, we can again think of filling the two squares. Again there are 100 choices for the first square. But what about the second square? In order for $n_1 + n_2$ to be even, n_2 must be even if n_1 is even *and* be off if n_2 must be even if n_1 is even *and* be odd if n_1 is odd. Thus in either instance there are 49 even or odd integers left, depending upon whether n_1 was even or odd. Hence there are $100 \cdot 49$ favorable outcomes.

(h) There are two cases to be considered for n_1: (1) n_1 is a multiple of 5 or (2) n_1 is not a multiple of 5.

Problem 5 What is the sample space in each case? We can think of counting the number of elements in the sample space, and the number of favorable outcomes as filling k squares, one for each digit of the k-digit numbers. Take $k = 4$.

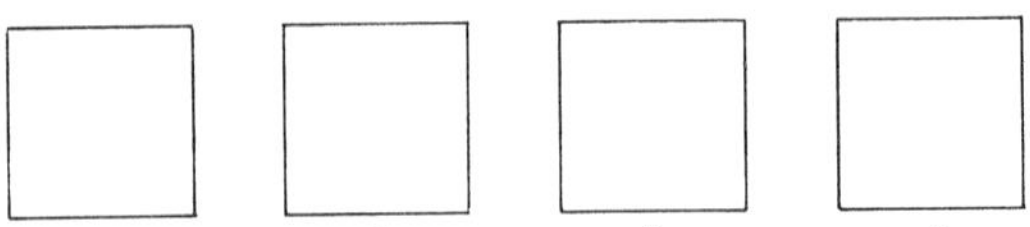

1 2 3 4

Then square 1 represents the thousand's place, square 2 the hundred's place, etc. How many choices are there for square 1 (it can't be zero!)? for square 2? etc. Keep in mind that in the sample space a digit can be used more than once, but not in the set of favorable outcomes.

Problems 7, 8, 9 See Chapter 6 to review the divisibility tests. In Problem 8 especially, keep in mind that each digit must be used once.

Problem 10 Consider the cases: a, b, c, d all even; a, b, c even, d odd; a, b even, c, d odd; etc.

18 Geometry

What do little acorns say when they grow up?

This chapter has a short introduction because you have done much of this material before in other contexts. The purpose of this chapter, then, is to examine geometric interpretations of some number concepts.

In order to make interpretations meaningful, I will insist that *all geometric figures considered have integral sides;* i.e. the lengths of all sides of any figure are integers.

The simplest geometric figures are triangles, so

Problem 1 Find all triangles whose perimeter is 13. (no units—inches, feet, yards, etc.—will be attached to any number. Just assume that the units are the same for any numbers mentioned in any problem.)

A *Pythagorean triangle* (x, y, z) is one whose sides x, y, z obey the Pythagorean Theorem, i.e. $x^2 + y^2 = z^2$; or, geometrically speaking, a Pythagorean triangle is a right triangle all of whose sides are integral.

Problem 2 Find the lengths of the sides of five Pythagorean triangles (x, y, z) where $\text{GCF}(x, y, z) = 1$.

Problem 3 Find all Pythagorean triangles (x, y, z) where x, y, z are three consecutive integers.

For each of the Problems 4, 5, 7, 8, 9 answer the question: To what number concept (or concepts) is this problem related? Try to find a problem in a previous chapter which is conceptually the same as each of the problems listed above.

Problem 4 If (x, y, z) is a Pythagorean triangle, prove that:

(a) x and y cannot both be odd.

(b) Either x or y is a multiple of 3.

(c) Either x or y is a multiple of 4.

(d) Either x, y, or z is a multiple of 5.

(It is quite possible for the same side to be a multiple of both 3 and 4: for example, (5, 12, 13) is a Pythagorean triangle.)

Problem 5 Find all Pythagorean triangles (x, y, z) if $y = 24$.

Problem 6 Three squares A, B, C are constructed so that the area of square B is 20 more than that of A, and 28 less than that of C. Find the length of the sides of all three squares.

Problem 7 Find x in the following figure if x is an integer.

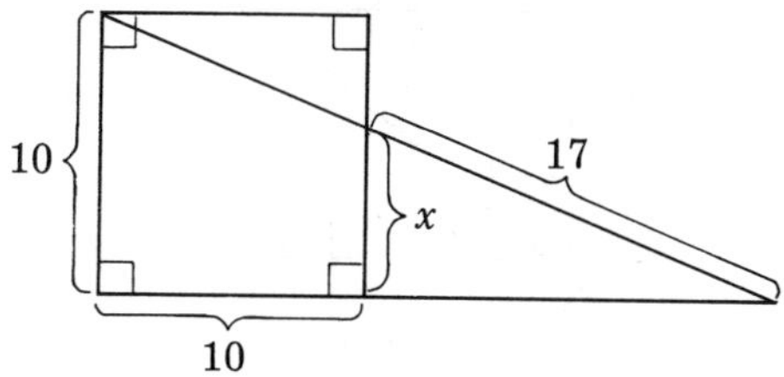

From triangles we now turn to 4-sided figures, in particular rectangles. Keep in mind that all sides of these rectangles have lengths which are integers.

Problem 8 Find all rectangles whose area is
(a) 17 (b) 30 (c) 72.
In general, how many rectangles are there whose area is n?

Problem 9 Find all rectangles each of whose area is numerically equal to
(a) its perimeter (b) twice its perimeter (c) three times its perimeter. Justify your answers.

Closing this chapter are two rather interesting problems.

Problem 10 Find five rectangles whose dimensions are chosen from the integers 1, 2, 3, . . . , 9, 10 (using each integer only once) and which can be put together to form a square. Draw a sketch of the square showing how the five rectangles are put together.

Problem 11 Consider a 4×5 rectangle subdivided into unit squares, as shown in the figure.

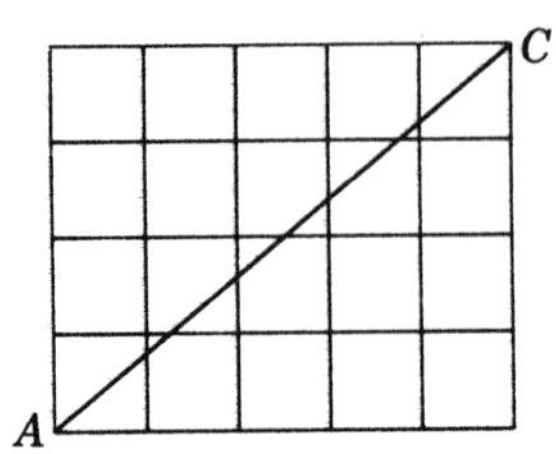

Draw diagonal AC. How many unit squares does the diagonal pass through? (The diagonal must pass through an interior point of a unit square; if it only passes through a vertex of a square, that square is *not* counted.

For example:  yes  no). Answer the same question for a

(a) 6×10 (b) 12×16 (c) 10×15 rectangle. Now determine a general formula for an $m \times n$ rectangle, and prove your formula if you can.

HINTS AND COMMENTS

Problems 4, 5, 7, 8, 9 You've done all of these problems before in other chapters. If you need help, page through the book for ideas.

Throw theory into the fire, it only spoils life.

— *M. Bakunen*

19 Some Famous (Unsolved) Problems of Number Theory

Numbers have long fascinated mathematicians probably because of their seeming simplicity. Yet many problems involving numbers have gone unsolved for centuries, despite the efforts of men with considerable talent. The solutions of some of those problems which have been solved seem to have an esthetic quality about them—a certain kind of beauty. Number theory is the essence of pure mathematics, and as such can be compared to any art form—music, literature, art, etc.—in which one admires the technique, the creativity, the originality of the artisan. (See Euclid's proof of the infinitude of primes.) But enough of this flowery prose.

We have previously encountered two of these classical unsolved problems:

(1) Do there exist infinitely many twin primes?

(2) Does there exist an odd perfect number?

Probably the most celebrated of the unsolved problems is:

Fermat's Last "Problem"

Does the equation $x^n + y^n = z^n$, $n \geq 3$, have any integral solutions? In other words, are there integers x, y, z such that $x^n + y^n = z^n$ for any $n \geq 3$?

We have previously encountered the equation $x^2 + y^2 = z^2$ in the Geometry chapter, and we saw that this equation had many solutions. In fact, there are formulas which determine all possible solutions to this equation. Regarding $n \geq 3$, Pierre Fermat (17 century French mathematician, regarded as the father of number theory) wrote in the margin of a book:

> However, it is impossible to write a cube as the sum of two cubes, a fourth power as the sum of two fourth powers, and, in general, any power beyond the second as the sum of two similar powers. For this I have discovered a truly wonderful proof, but the margin is too small to contain it.

Though several prizes have been offered for this proof, and many thousands of people, not only mathematicians, have endeavored to find it, the "truly wonderful proof" has remained elusive. It is thought by many mathematicians that Fermat's "proof" was false, for despite the many advances in mathematical knowledge since his time, no proof has yet been discovered.

Below are a sampling of problems involving some simpler cases of Fermat's Last "Problem."

Problem 1

(a) Prove that $x^3 + y^3 = z^3$ has no solutions where (x, y, z) are consecutive integers.

(b) Prove that, in general, $x^n + y^n = z^n$ has no solutions where (x, y, z) are consecutive integers.

Problem 2 Show that $x^n + y^n = z^n$, $x < y$, has no solutions if $x = 1$.

Problem 3 Show that $x^6 + 2^6 = z^6$ has no solutions with $\text{GCF}(x, z) = 1$.

Problem 4 Prove that $x^n + y^n = z^n$ has no solutions if $x < n$ and $y < n$. Try at least one special case first.

Problem 5 See if you can state and solve another special case of Fermat's Last "Problem."

The nature of a prime integer is often mystifying. The facts that

(a) there are infinitely many primes (Euclid) and

(b) there are arbitrarily large gaps in the set of primes (Problem 7, Chapter 14, page 103)

are seemingly contradictory. An early problem was to find a formula that produces only primes, i.e. a function $f(x)$ such that for every integer n, $f(n)$ is a prime. The natural place to begin one's search is with polynomials.

Problem 6 Find a quadratic polynomial $P(x)$ such that $P(0) = 2$, $P(1) = 3$ and $P(2) = 5$. Is $P(n)$ always prime?

Problem 7 Consider the polynomial $P(x) = x^2 - x + 41$. Compute $P(1)$, $P(2)$, $P(3)$, $P(4)$, $P(5)$, $P(40)$. Is $P(n)$ always prime?

Problem 8 Show that no polynomial $P(x) = a_0 + a_1 x + a_2 x^2 + \cdots + a_n x^n$ with integer coefficients, where $a_0 > 1$, can represent a prime for every integral value of x.

In 1948, W. H. Mills proved the striking result that there exists a non-integral real number θ such that the greatest integer less than θ^{3^n} is a prime for all values of n.

As a final example of a famous unsolved problem, we state

Goldbach's Conjecture

Conjecture 1 Every even integer can be represented as the sum of two odd primes.

Some other conjectures related to Goldbach's Conjecture are:

Conjecture 2 Every odd integer can be represented as the sum of three odd primes.

Conjecture 3 Every even integer greater than 4 can be represented as the sum of three primes.

Problem 9 Write 16, 20, 24, and 60 as
(a) the sum of two odd primes in as many ways as possible;
(b) the sum of three primes in as many ways as possible.

Problem 10 Write 17, 21, 25, 61 as the sum of three odd primes in as many ways as possible.

Problem 11 Show that Conjecture 1 and Conjecture 3 are equivalent; that is, if Conjecture 1 is true, then Conjecture 3 is true; and if Conjecture 3 is true, then Conjecture 1 is true.

Problem 12 Show that if Conjecture 1 is true, then Conjecture 2 is true.

Conjecture 2 was proved in 1937 to be true for all sufficiently large integers, while the most recent result with regard to Conjecture 1, announced in 1966, is that every even integer is the sum of a prime and an integer with at most two prime factors.

These problems indicate that number theory, much of which was discovered by the ancient Greeks, remains today as one of the most baffling areas of mathematics in that many problems which are so easy to state are so difficult to solve. But I guess that is just the nature of our universe.

An excellent book which contains descriptions of many unsolved problems in all areas of mathematics is *Tomorrow's Math,* C. S. Ogilvy (Oxford University Press, 1972).

PROBLEM SOLVING
FROM THE TEACHER'S POINT OF VIEW

This last section is primarily for students who are, or plan to be, teachers. I would like to discuss three questions concerning problem solving in the classroom:

(1) Why should problem solving be taught?
(2) What is the teacher's role in teaching problem solving?
(3) What are some specific suggestions for incorporating problem solving into the classroom?

Here are my answers. After reading this discussion, you should formulate your own answers to these questions, i.e. you should begin to answer the general question "What role should problem solving play in *my* classroom?"

(1) I repeat a portion of the introduction:

> A teacher of mathematics has a great opportunity. If he fills his allotted time with drilling his students in routine operations he kills their interest, hampers their intellectual development, misuses his opportunity. But if he challenges the curiosity of his students by setting them problems proportionate to their knowledge, and helps them to solve their problems with stimulating questions, he may give them a taste for, and some means of, independent thinking.

Thus one reason for teaching problem solving is to develop the ability to think independently. Solving problems gives a student a sense of accomplishment and eventually develops self-confidence toward mathematics.

Another equally important reason is to develop the ability to recognize patterns, to make "educated" guesses. Faced with a problem similar to those in this book, the first step in solving it is to compute several numerical examples and then to make a guess, a conjecture, based on these examples. Most important discoveries (in any area, not just mathematics) began as guesses or conjectures; this is, after all, how the scientist thinks. He makes a conjecture first, and then tries to verify it; the guessing stage in any discovery is at least as important as the verification stage.

Finally, posing problems is one of the few ways we can communicate to our students that mathematics is an "alive" subject—that there are still many unsolved problems, many discoveries yet to be made. With conventional teaching methods, students can get the idea that mathematics is a closed subject, and thus lose interest in it. Loss of interest is one of the major reasons, I feel, why many students don't like mathematics.

Suffice it to say I am strongly in favor of teaching problem solving in the classroom. I realize that problem solving is a slower method of teaching, not as much material can be "covered" in a given amount of time. Nevertheless there is, in fact there must be, a certain amount of time which can be allotted to this means of teaching—its benefits are too important to be neglected.

(2) Since there is no sure-fire method of solving a given problem, it follows that there is no sure-fire method to teach problem solving. Just as each student develops his own style of solving problems, each teacher develops his own style of teaching problem solving. The best I can do here is to offer a few general observations:

a) The most important aspect of teaching problem solving is to arouse the curiosity of the student—make the student *want* to solve the problem. As I've said before, desire is 50% of performance.

b) His curiosity aroused, your task, then, is to give the student "just enough" help. After a student has worked on a problem for a time without success, frustration sets in. You must see that this frustration does not become "critical." A certain amount of frustration can be beneficial, it makes the final success all the more satisfying. To persevere despite failure, to make progress toward the essential idea, and finally to conceive the solution can certainly build the self-confidence of anyone. But generally a student needs some help. Check over his work for errors, discuss his progress so far and what his difficulty is (if he is able to verbalize his difficulty, he can frequently overcome it), give him a hint to "get over the hump"; in general, make him aware of the suggestions and faults we discussed in the first six chapters.

c) Give the student immediate reinforcement, as immediate as is practically possible. Compliment him for those problems he solved correctly; point out his errors for those he solved incorrectly. Encourage him to redo these problems. Ideally, you will have done your best possible teaching job when the student no longer needs you—when he *can* think independently.

d) Polya reminds us of Anatole France's comment on an essential quality possessed by a good teacher of problem solving;

> Do not try to satisfy your vanity by teaching them a great many things. Awake their curiosity. It is enough to open the minds, do not overload them. Put there just a spark. If there is some good inflammable stuff, it will catch fire.

I must confess that I have, on occasion, disregarded Anatole's advice, by teaching my students a lot of stuff simply because I wanted to let myself know that I understood it.

(3) Below is a list of specific comments and suggestions regarding problem solving in the classroom.

a) In solving any problem, you are practicing computing skills in constructing your examples. Thus asking your students to solve a problem is asking them to practice computation. This approach to practicing computation seems far more satisfying than the drudgery of drill exercises. One example: In completing the magic square

	6	
7	4	

the student performs several additions and subtractions as he does in the following problem: Take any two integers, find their sum and (non-negative) difference. Then find the sum of that sum and difference. How is this last sum related to the original two integers?

b) Solving problems is also an excellent way of probing deeply into basic concepts. The student who successfully solves the problem of finding all integers whose remainder upon division into 100 is 1, gains a great insight into the meaning of the Division Algorithm.

c) When discussing a certain concept in class try to get the class members to help you formulate problems involving that concept. When the student participates in the formulation of the problem, he has added incentive for solving it.

d) Having students solve problems in class is one method to involve every student, something a lecture cannot hope to do. Solving problems in groups can also be effective if everyone in the group makes a contribution. Working among themselves, students will more readily question an idea put forth by another student than one put forth by a teacher because of the "teacher is always right" syndrome.

e) Reverse the usual procedure in a problem solving situation. Give the students an answer (e.g. 13), and ask them to formulate several problems having this answer.

f) You can incorporate problem solving into your class in several ways: (i) Supplement the textbook drill exercises with your own problems (many of which can be constructed by making problems out of statements in the text or formulating a problem based on the text exercises); (ii) Post a "problem of the week" on the bulletin board; (iii) Keep a "creativity file," a set of problems on index cards. A student chooses a problem which interests him, copies it, and signs his name on the back of the card before replacing it; (iv) Organize an after-school mathematics competition among students in your school or between schools. Solving two or three problems in a specific time limit, either individually or in groups, can be stimulating. Many states offer state-wide mathematical competitions of this sort.

g) Some other topics related to problem solving in the classroom are: (i) the connection between mental exercises in arithmetic and problem solving, (ii) solving problems related to practical applications of arithmetic, (iii) certain simple problems can be solved by concrete manipulations with physical objects (e.g. Cuisenaire rods for adding rational numbers).

Finally, I think you should now take some time to practice constructing problems. Choose a grade level (preferably one you wish to, or are currently, teaching). Construct several thoughtful problems suitable for your grade level (you can base them on any problems contained in this book). With regard to each problem, present your solution to the problem and then discuss the following questions:

(1) How much previous knowledge is needed to solve the problem?
(2) What mathematical idea(s) are illustrated by the problem?
(3) What mathematical idea(s) (if any) may be motivated by the problem?
(4) What hint(s) would you give for the problem?
(5) Any other comments you want to make (e.g. can you make a game out of the problem? is the problem more suitable for slow or fast learners? does the problem have a practical application? etc.).

Keep in mind that your major objective here is to construct problems which, in your opinion, require some thought (i.e. cannot be done by rote memory) *by the student for whom they were intended.* (They may seem trivial to you.)

Teaching mathematics by problem solving is hard work. You must pose the problems, answer questions (sometimes immediately—many times a day later after you've had time to think it over), give appropriate hints and encouragement, evaluate solutions and attempts at solutions, and occasionally wage a psychological battle with a student to get him to perform up to his potential.

But the rewards are worth the effort. You will learn more mathematics than your students, you will teach your students to think, and you will stay mentally alive and alert so that your teaching from year to year will really be a challenge instead of a chore.

The Final Problem

There is only one way to end this book:

Problem What is the next letter in the sequence

O, T, T, F, F, S, S, E, N, ?

Index

1 2 3 4 5 6 7 8 9 10 11 12 13 14 15 16 17 18 19 20 21 22 23 24 25 80 79 78 77 76 75 74 73 72